AF598144

Praise for

BIONIC RETAIL

"Gary has done it again! *Bionic Retail* is a must-read for leadership in the supermarket industry. Today's supermarket experience continues to evolve, and Gary has done an excellent job of explaining the speed of technology and how it will impact the supermarket experience in the future. Supermarket leaders should be focused on the speed of retail change and how those technological advancements can positively impact the customer experience in the future."

STEVEN C. SMITH, President and CEO,
K-VA-T Food Stores, Inc.

"In his latest book, Gary Hawkins leverages his extensive experience to provide unique insights on how tech-driven disruption is shaking up the retail sector. A very useful read for retailers and other commentators alike, and in fact, for anyone thinking about how newer technologies (e.g. generative AI) will impact retail. Now more than ever, retailers need to embrace change and transformation, and this book will help them on that journey."

RICHARD CHAMBERLAIN, Managing Director,
General Retail, RBC Capital Markets

"Amidst the ever-changing landscape of the grocery industry, adopting the rapid technological advancements is not merely an option; it's the gateway to a future where convenience, sustainability, and innovation harmonize to nourish both our bodies and the world we inhabit. Gary's book *Bionic Retail* underscores the importance of cultivating the mindset and culture within the grocery industry to embrace the swift pace of technology innovation."

DOUG BAKER, Vice President of Industry Relations, FMI, The Food Industry Association

"We have come to discover that courage and curiosity are two of the most critical leadership traits. Gary Hawkins expands on these character traits by identifying the outcomes: inspired vision, committed action, and an innovation doctrine. Hawkins is a creative storyteller who is passionate about helping retail industry leaders learn from his lessons and the lessons of others. In his most recent book, *Bionic Retail*, Hawkins uses intentional metaphors, historical perspectives, and interesting new facts about technology to enlighten future leaders. *Bionic Retail* is a must-read for every leader in the retail and CPG industries who is wondering how to compete and win in the future. I encourage all my students to read Hawkins' books so they are prepared to compete in a world where AI will drive our strategic decisions and enhance our ability to innovate like never before."

DR. CYNTHIA MCCLOUD, Director and Adjunct Professor, USC Marshall School of Business Food Industry Programs

"Another excellent book from Gary Hawkins. *Bionic Retail* is a must-read for today's grocery industry leaders at all levels. It clearly articulates what is happening in today's world with impactful examples from the grocery industry of yesterday and today. After reading *Bionic Retail*, I have a much clearer understanding of the speed of change and the 'danger of doing nothing' or worse, 'doing what we have always done.'"

MIKE STIGERS, President, Wakefern Food Corporation

"*Bionic Retail* by Gary Hawkins is an absolute must-read for anyone in the world of retail, from seasoned veterans to aspiring leaders. Hawkins delivers a compelling wake-up call, highlighting the urgent need for transformation in the face of exponentially accelerating innovation in the industry. The book is a treasure trove of real-world examples that pull back the curtains behind some of today's most significant innovations, providing invaluable insights and inspiration. It not only illuminates the path toward cultivating a tech-forward culture but also offers a pragmatic framework for getting started and building the organizational muscle. *Bionic Retail* is the resounding call to action that laggards require and a resounding endorsement for those who have leaned into transformation with enthusiasm. Gary Hawkins has crafted a masterful guide to the future of retail that should not be missed."

TOM FURPHY, CEO, Consumer Equity Partners

"Gary Hawkins has been a leading voice in the world of retail technology, and with *Bionic Retail*, Gary is sharing his expertise with retailers at a critical moment in time when technology is changing business faster than we can adapt. This book is a must-read for any retailer or business leader looking to develop a customer-focused strategy to prepare themselves for the exponential world into which we are entering."

KARTHIK EASWAR, Associate Teaching Professor of Marketing, McDonough School of Business, Georgetown University

"This is Gary Hawkins' most ambitious and important book to date. *Bionic Retail* is a consilience of sorts, a unity of knowledge, combining technological innovation with human inspiration, as applied to the consumer goods and food retailing industries. Inventive technology continues to accelerate at an exponential pace, widening the gap between functionality and utility. Those companies that move closer to the former through pipelines, applications, and platforms will enjoy a growing competitive advantage. The laggards will be at risk. Gary deftly lays out the case for cultural and organizational change and preparing for a new technology-enabled world that couldn't be imagined—much less lived in—not too long ago. This is a must-read for all C-suite food industry executives."

MARK W. BAUM, Chief Collaboration Officer, FMI, The Food Industry Association

"Gary Hawkins' latest book, *Bionic Retail*, is a must-read for anyone in the retail industry. It's a thought-provoking exploration of the future of retail, where physical automation and artificial intelligence are integrated through a digital nervous system into our human organizations to create a new retail species: *Bionic Retail.* Hawkins' deep experiences with retailers and technology providers shines through, providing a blueprint for navigating the rapidly changing retail landscape. This book challenges traditional retail operating and business models, urging retailers to look beyond the shiny new innovations and focus on aligning their organizational structures and decision-making processes with the future. *Bionic Retail* is more than just a book; it's a road map to the future of retail."

SHISH SHRIDHAR, Global Retail Lead, Microsoft for Startups

"Gary Hawkins manages to quote Ray Bradbury and Lewis Carroll, cite *Star Trek* and *The Six Million Dollar Man*, and use his Bernese Mountain Dog puppy as a business metaphor in *Bionic Retail*, which makes him my kind of technology writer: accessible and relevant, making a persuasive case for how to actually act on actionable information."

KEVIN COUPE, "Content Guy," MorningNewsBeat.com

BIONIC RETAIL

www.amplifypublishing.com

Bionic Retail: How to Thrive in an Exponential World

For more information, please contact:
Amplify Publishing, an imprint of Amplify Publishing Group
620 Herndon Parkway, Suite 220
Herndon, VA 20170
info@amplifypublishing.com

Library of Congress Control Number: 2023919776

CPSIA Code: PRV1123A

ISBN-13: 979-8-89138-063-9

Printed in the United States

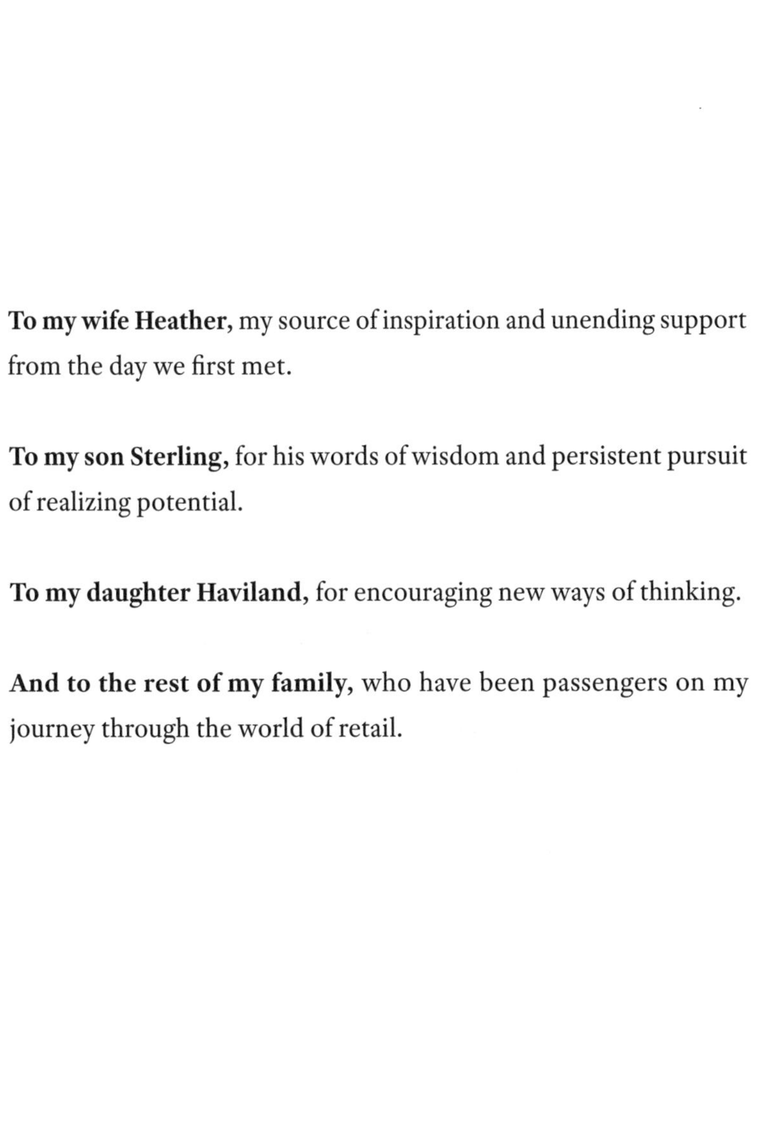

To my wife Heather, my source of inspiration and unending support from the day we first met.

To my son Sterling, for his words of wisdom and persistent pursuit of realizing potential.

To my daughter Haviland, for encouraging new ways of thinking.

And to the rest of my family, who have been passengers on my journey through the world of retail.

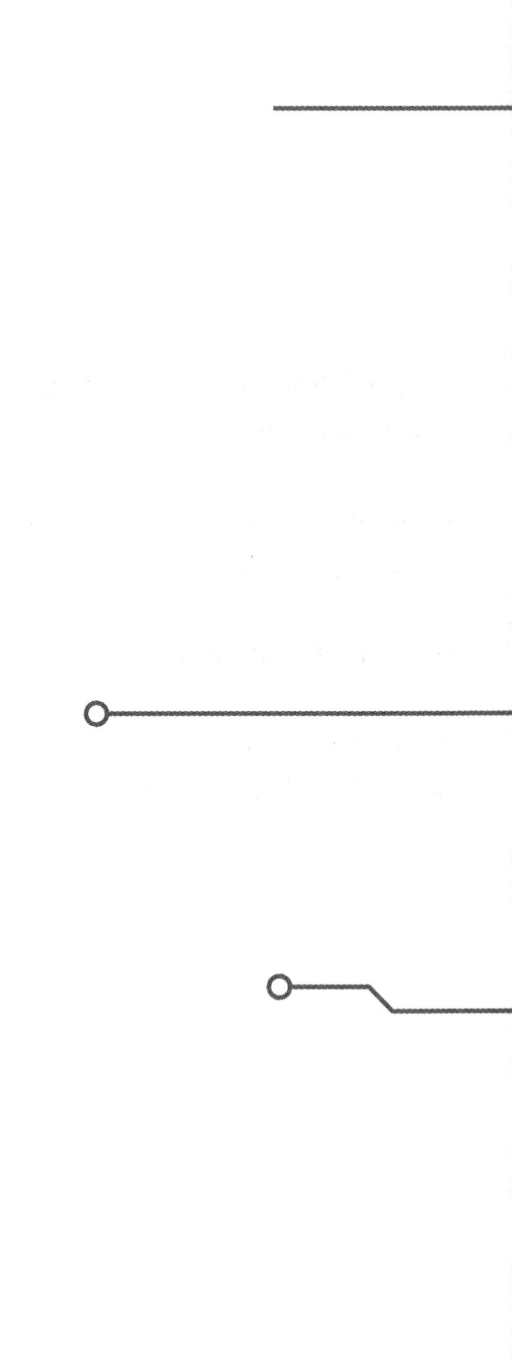

GARY HAWKINS

BIONIC RETAIL

HOW TO THRIVE IN AN EXPONENTIAL WORLD

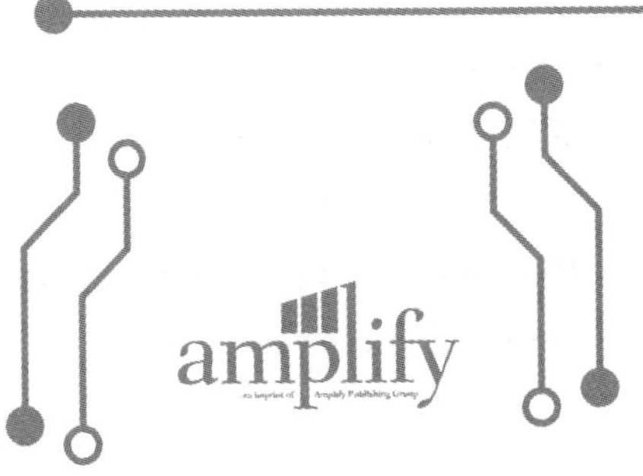

CONTENTS

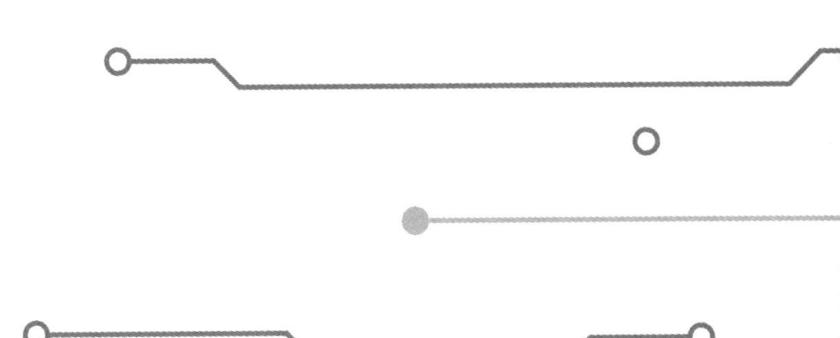

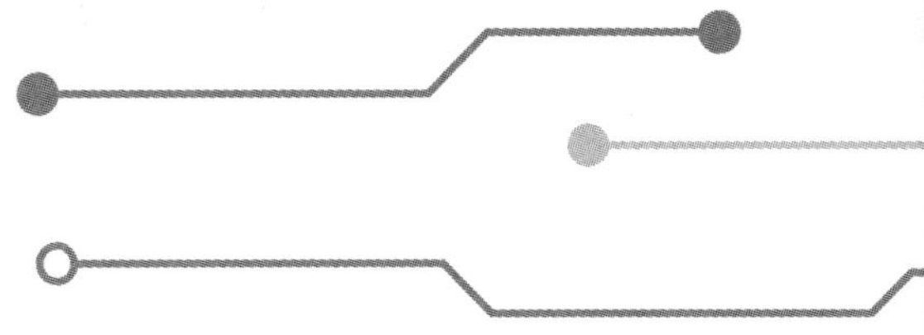

INTRODUCTION

Everything had come to a halt in one of the worst blizzards to strike the Syracuse, New York, area in decades. In what became known as the "Storm of the Century," 43 inches of snow had fallen in the span of about thirty-six hours. It was 5:00 a.m. on Sunday, March 14, 1993, and the snow-filled streets offered a foreboding start to what would come to be the most pivotal day of my business life.

The journey had started a couple years earlier when I began reading about a few retailers in Asia who had launched frequent shopper programs, giving their shoppers a barcoded card to use when they shopped that triggered the delivery of electronic coupons. Until this revolutionary approach, it was common practice for retailers to require shoppers to tear out coupons from the weekly ad to get the hottest sale prices. It was the only practical way for retailers to limit the sale of loss-leading ad items.

But as I thought more about what was happening, it became clear that the unique barcode on the frequent shopper card could be embedded into the transaction, not just trigger an electronic coupon. By attaching the unique barcode number to the transaction, it would enable a retailer to build a database of shopper-identified transactions. And from that data, retailers would be able to glean new insights. Little did I know where that idea would lead.

At the time, I was running our family's supermarket. The store had started as a summer farm stand in 1934 selling the surplus vegetables from my great-grandmother's victory garden. And while the store was successful, we were facing growing competition from regional retailers like P&C, Price Chopper, and Wegman's, perhaps one of the best food retailers in the United States. Walmart was entering grocery retail, already impacting traditional retailers in markets where they opened their original supercenters. Competition was going to get even tougher.

As a single store operator, I had limited resources to work with as I pursued my frequent shopper vision. I spent time researching and gathering cost estimates to build the database needed to support such a program, quickly concluding it was going to be too costly to develop and operate.

But I refused to let go of the vision of what could be. A few months later, I attended the 1992 FMI Show, the major industry convention that drew tens of thousands of retailers from around the world. It was there I encountered Don Irion, who, with his partner John Schultz, had started a company called Retail Marketing Systems, or RMS. Their product was MarketExpert, a system designed to support retail frequent shopper programs.

What Irion and Schultz had built was just what I had envisioned was needed to launch and operate a frequent shopper program. We needed the ability to create a customer record of name, address, email, etc. and attach it to a frequent shopper number, along with the ability to build a database of shopper-identified transactions and then query the data, creating customer segments and lists. This would allow us to target promotions to a segment of customers. And more. This technology enabled a powerful strategy for attracting and retaining customers in retail.

My family's supermarket quickly became RMS's first client and planning for launching our frequent shopper program began in earnest. Challenges began to arise almost immediately. Technical obstacles appeared early on—specifically the need to modify the point of sale (POS) system to accept and embed the customer's loyalty number into the transaction and use that identifying number to trigger discounts. We had to develop all new signage to inform shoppers that obtaining the sale price required the use of their frequent shopper card.

In retrospect, the most significant hurdle to pursuing the vision was people. Our management team quickly let me know they thought I was nuts. They believed there was no way customers would sign up, provide their private information, and then carry a card that had to be used to obtain sale items and all the other discounts. And it wasn't just our leadership team who expressed doubts. Our wholesaler also was concerned that what we were planning would hurt business. Vendors quietly voiced their fears. They thought we were adding challenges to getting sale items, not making it easier.

But I was inspired by the vision that had taken hold in my head, that customer data would enable us to change the way we went to market, reducing reliance on the traditional hook of low prices to drive business. Time was invested to help our team understand what the few other retailers at the time were doing, how the technology worked, and why I believed that this path would help ensure our future. Butting heads with Walmart on low prices was not the way to success.

In retrospect, what I've come to appreciate is that the pushback was caused by certain beliefs our management team, associates, and vendors held around how they thought retail should be done. I see the retail industry confronting this same challenge today with the rise of AI and machine learning: long-established processes and beliefs frequently constrain innovation. It is only by identifying those constraints and setting them aside can progress be made. Except this time, progress is essential to survival.

So, in the middle of the worst blizzard to hit the Syracuse area in decades, we launched our loyalty program. Within weeks, we were identifying over 80 percent of our sales to shoppers. The journey had begun. Along the way I've encountered, and blown up, countless practices that had become engrained in retail, from the weekly ad to transforming management and financial reporting by including customer-based metrics, and many more.

Through this time, technology was becoming increasingly important. The growing consumer use of the World Wide Web and email helped trigger the next inspiration: marketing personalization.

Dissecting detailed shopper-identified transaction data had given me a never-before-possible view into customer-level economics and the impact of traditional marketing like the weekly ad. I added to this knowledge a deep dive into consumer packaged goods (CPG) brand marketing, understanding the vast waste in CPG marketing funds. There had to be a better way.

While RMS's MarketExpert system enabled us to target promotions to customer segments, the laborious process to do any kind of personalization even at the customer-segment level was incomprehensible. We needed a new tool.

In 2001, the inspiration to provide different shoppers with relevant promotions and personalized savings led to a startup called Seventh Street Software out of Berkeley, California. They had developed the first targeting engine using Bayesian statistical modeling to automatically generate relevant offers for an individual shopper based on purchase history.

What Jesse Quatse and his partner C.V. Ravi had developed was brilliant. But it was like a Porsche engine with no surrounding vehicle or road to run it on. More was needed to make it actionable.

As I'll relate later in the book, a classic middle-of-the-night epiphany gave rise to what we came to call SmartShop, the first true marketing personalization system for retail. SmartShop was a complete personalized marketing solution with Seventh Street's automated engine at its core, that enabled retailers to create an offer pool, personalize the offers, and then communicate to shoppers via email or website. That work led to combining SmartShop with biometric authentication (a finger scan and PIN number) for discount savings and payment in 2005, nearly a decade before Apple Pay.

The key takeaway from relating my journey is the power of inspired vision. Little did I know that launching one of the first frequent shopper programs in the United States would lead me on a life-long journey of international speaking, becoming a strategic advisor to prominent retailers, CPG brand manufacturers, and technology companies on five continents, a technology innovator and thought leader, and an author. Stepping away from being an active retailer over a decade ago, I moved from using our store as a live lab to test new tech, to forming, with my son Sterling and my daughter Haviland, CART, the Center for Advancing Retail & Technology, to share our knowledge and experience with the broader retail industry.

That's one of the most powerful things about inspired inspiration. It inevitably opens other doors and creates unforeseen possibilities and opportunities. And that's important as we continue discussing creating the future of retail.

The Importance of Inspiration

A decade-old *Harvard Business Review* article calls out the importance of inspiration. "In a culture obsessed with measuring talent and ability, we often overlook the important role of inspiration. Inspiration awakens us to new possibilities by allowing us to transcend our ordinary experiences and limitations. Inspiration propels a person from apathy to possibility, and transforms the way we perceive our own capabilities."* The article goes on to call out three main qualities of inspiration: evocation, transcendence, and approach motivation.

"First, inspiration is evoked spontaneously without intention. Inspiration is also transcendent of our more animalistic and self-serving concerns and limitations. Such transcendence often involves a

---

* Scott Barry Kaufman, "Why Inspiration Matters," *Harvard Business Review*, November 8, 2011, https://hbr.org/2011/11/why-inspiration-matters.

moment of clarity and awareness of new possibilities. This moment of clarity is often vivid, and can take the form of a grand vision, or a 'seeing' of something one has not seen before. Finally, inspiration involves motivation, in which the individual strives to transmit, express, or actualize a new idea or vision. Inspiration involves both being inspired by something and acting on that inspiration."*

I'm going to add one more quality of inspiration: the power of *sharing* it.

One of my clients years ago was M&M Meat Shops, a unique food retailer in Canada. Mac Voisin, the owner and CEO, was inspired by the vision to recreate retail by building customer relationships and leveraging data-driven insights. As part of sharing my vision at their annual convention, I had my son Berkeley, who was around fourteen years old at the time, dress up as a turkey and come on stage to highlight the story I told of running the first free Thanksgiving turkey program through our nascent loyalty program in 1994. While Berkeley may still be struggling with some emotional trauma from being a turkey on stage, the presentation left an indelible impression on the M&M team.

In similar fashion, my vision to use customer data to transform the way retail was done would have been worthless had I not taken action, shared it with our team and our customers, and worked to make it a reality. Too often do people have great ideas but fail to take action. There is perhaps no greater failure of potential than failing to pursue a true and powerful inspired vision.

---

* Ibid.

A Blueprint for the Future

As we look to how the retail industry is grappling with accelerating technology innovation, we increasingly see that technology itself is not the problem. Instead, as the pace of tech-driven change outside the organization grows faster than the pace of change within the organization, there is a growing misalignment. And that misalignment is rooted in the culture and organizational structure itself. So, what's needed is a plan, a methodology, to guide retailers into the future. And that methodology is based on three things:

1. **Developing an inspired vision.** Without a compelling vision of what retail's future looks like, retailers will be destined to continue implementing technology without a grand plan, a scatter-shot approach that chews up resources while creating less and less benefit.
2. **Committed action.** No matter how inspirational the vision, it is meaningless without taking committed action to make it happen. And yet we see retailers time and again failing to commit to purposefully creating their future. Rather, too many retailers continue to propagate past practices that contribute to the growing gap between the linear path to purgatory and the exponential path to the future.
3. **Creating—and living by—an innovation doctrine.** An innovation doctrine establishes the principles, strategies, and guidelines that help govern the organization and provides a common understanding. This doctrine needs to become embedded in your organization or you risk your company's "immune system" kicking in to reject new innovation and new practices.

A Unique Point in Time

Think about the world we live in today. New, exciting innovations appear almost daily in every aspect of our lives. There are self-driving cars. Nearly any product you can imagine is available on your doorstep within hours. Medical advances that were the subject of science fiction just yesterday are being used to help patients right now. You can ask your smartphone pretty much any question and receive an answer in less than a second. In every facet of our lives—the ways we communicate, how we work, the things we buy, how we travel, and more—innovation is exploding. Every day, we're introduced to something new.

Think about all the new capabilities flooding into retail . . . the ability to just walk out of a store with the products you've picked up; payment as easy as waving your hand over a terminal; orders fulfilled by an automated system, delivered by a robot. Things that seemed like another world just a few short years ago. Shopping is transforming in real time. Customers are receiving guidance to specific foods across the store beneficial to their individual health conditions. Product offers and savings magically appear on shoppers' smartphones based upon their real-time location in the store. Products are delivered to the home by self-driving vehicles or drones.

And these things are here *today*. What's coming next includes bringing the physical store to life through capabilities like next-generation smart glasses and augmented reality. Consumers are gaining control and ownership of their data, blowing up retail media networks and disrupting the business models of so many companies. They will interact with advanced computer systems wherever they are, through voice, and even gestures.

We are at a unique point in history where a discussion, understanding, and action, focused on creating inspired visions, is of paramount importance. As we'll read about in upcoming chapters, we have reached and surpassed the inflection point on the exponential growth curve of technology and have entered a world unlike any before. We can remain in our comfort zones, defending past practices and complaining about the complexity and costs of change. Or we can open to the new possibilities and opportunities that accelerating and burgeoning innovation provide to create something that has not before existed. We can create Bionic Retail.™

Bionic Retail™ fuses technological capability with human knowledge and experience to make retail better, stronger, and faster. The "glue" is human creativity, which enables each retailer to pursue and realize their own vision of retail's future while providing an opportunity to regain retail's customer heritage.

No "Easy Button"

Too many retailers look for the Innovation Easy Button, believing that by just plugging the latest technologies into their existing, decades-old processes, they will be successful. But there's more to do than just transplant some technologies into your organization. Just like in the human body, transplanted ideas and approaches are at risk of rejection by your business if they are not introduced correctly. As we zoom past the inflection point on technology's exponential growth curve, traditional retailers must transform their organizations to lessen the aversion to change that can accompany new innovation, undergoing a metamorphosis, emerging with the same

DNA but a very different structure adapted to the new environment.

So, there is work to be done. As retail executives learn about and understand this new world we now inhabit—where tech-fueled change continuously accelerates—onboarding, socializing, building, and implementing these new approaches needs to take place. Just because the retail experience is becoming more digital doesn't mean successfully executing these changes won't take real work from real people, we're just all going to have to think differently to succeed. Alongside that, we see a growing dichotomy between what traditional retailers believe is possible and what digital natives with a tech-first mindset know can be done and what will likely come to our reality in retail very soon.

The fast-moving consumer goods retail industry—ironically—has historically been slow moving, any innovation was driven from within, inevitably focused on cost reduction and gaining efficiency. There is a long track record of resistance to new capabilities: Merchandisers refusing to believe AI systems can plan weekly ads better than their years of experience. Management refusing to manage by customer metrics, maintaining yesterday's product focus. Marketers digitizing the decades-old mass-promotion-filled weekly ad, thumbing their noses at the proven power of personalization. The list goes on and on.

Results from traditional retailers will not change until their beliefs do. Changing beliefs opens the door to new possibilities, and new possibilities are the key to creating the future of retail.

Danger, Will Robinson!

Some may remember the old television series *Lost in Space* that followed the adventures of the Robinson family as they travel the galaxy trying to get back home. One of the key characters is Robot, who inevitably warns the young Will Robinson that danger lies ahead.

Much like the Robinsons, whether we've willingly joined the journey or not, we're all navigating the very unknown galaxy of retail's future. There are dangerous times ahead for retailers, and retailers who refuse to release their hold on traditional models are the most at risk. As technology-fueled innovation continues to explode, new capabilities, new digital-native competitors, and new business models will proliferate. Risks to traditional retailers will grow in lockstep with the exponential growth of technology.

"Since 2000, 52 percent of companies in the Fortune 500 have either gone bankrupt, been acquired or ceased to exist. U.S. corporations in the S&P 500 in 1958 remained in the index for an average of sixty-one years. By 1980, the average tenure of an S&P 500 firm was twenty-five years, and by 2011 that average shortened to eighteen years based on seven-year rolling averages. These are challenging times for companies as the speed, volume and complexity of change intensify."*

And disruptive change is not spread evenly: a study by the U.S. Department of Agriculture's Economic Research Service found that independent grocers—defined as operators with four stores or fewer—saw their market share decline in 44 percent of U.S. counties

---

* Capgemini Consulting, "When Digital Disruption Strikes: How Can Incumbents Respond?" 2015, https://www.capgemini.com/consulting/wp-content/uploads/sites/30/2017/07/digital_disruption_1.pdf.

between 2005 and 2015.*

Tech-driven disruption is playing an increasing role in this shake-up as retailers are forced to keep pace with fast-changing shopper expectations. And the pace of technology adoption is picking up. Statista reports that it took mobile phones twelve years to reach 50 million users. Facebook took four years to reach the same scale. Pokemon Go, the wildly successful augmented reality game app, took nineteen days to reach 50 million users.†

The pandemic provided the retail industry a warm-up for the accelerating change that is upon us. Almost overnight retailers were forced to deploy or massively scale up their online shopping capabilities. Sadly, the lessons learned around speed and nimbleness seem to have been quickly forgotten as many of those same retailers have settled back into their slow-moving ruts as the crisis subsided.

Intentions

If anyone tells you they have the answers to succeed in the time ahead—they don't. The world we have now entered has never existed before. We are in a world where technology—particularly artificial intelligence—is accelerating, changing jobs, changing work, and

---

* Jeff Wells, "Independent Stores Are Losing the Battle against Chain Operators, Study Says," Grocery Dive, November 30, 2017, https://www.grocerydive.com/news/grocery--independent-stores-are-losing-the-battle-against-chain-operators-study-says/534450/.

† Patrick Wagner, "The Road to Ubiquity Is Getting Shorter," June 22, 2018, https://www.statista.com/chart/14395/time-innovations-needed-for-50-million-users/.

transforming how we interact with the world around us.

While caterpillars are genetically programmed to spin a chrysalis and begin their transformation, humans are different. We need inspiration to change; some kind of catalyst. Without that inspiration, retailers will simply be dragged along for the ride as the world continues to evolve. Slower moving retailers will increasingly fall further behind competitors leveraging technology and embracing it to create a completely unique shopping experience for their customers. Inspiration, especially inspiration around a vision of what future retail can be, becomes our own inflection point, moving us from the linear path we have lived our lives on to the exponential path in sync with advancing technology.

My intentions with this book are simple: To share with readers a blueprint for how to navigate the uncharted waters we've now entered. A system built on identifying the beliefs and practices that stand in the way of new innovation, developing an inspired vision, and committed action to create the future.

But I can promise you that realizing that potential future is not easy. As I said earlier, there is some work to do. Traditional retail has little time to recalibrate and adjust to the exponential world in which it now operates. That's what this book is about. Helping retail leaders understand this new world and the rules that govern it. And while there are challenges that await around every corner, there is hope.

I'm not going to varnish the hard truths that the retail industry must confront, and they must confront them starting now if they want to succeed. That's been the biggest problem: retail leaders failing to understand the implications of the exponential world in which we now live. Many were lulled into complacency by strong pandemic sales and more recent inflation-driven sales and margins. What many must realize is that a historically risk-averse industry is actually becoming more at risk as the world transforms ever-faster. Not taking risks actually increases the risk of failure.

But beyond conveying that wake-up call, I am more excited than I have ever been about how retail can transform and reinvent itself to serve the customer. Technology itself can help retail transform and thrive in this new world of ever-accelerating change. If you can realign your leadership and organizations to the demands of this new world, there is a massive opportunity on the other side.

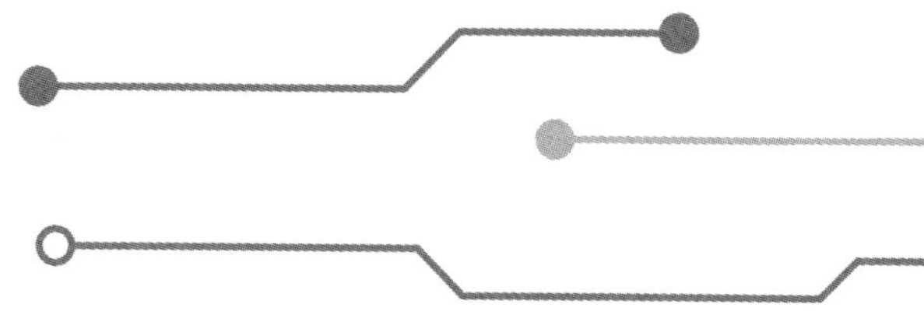

Part 1

THE WORLD PAST THE INFLECTION POINT

"If we wish to make a new world, we have the material ready. The first one, too, was made out of chaos."

—Robert Quillen

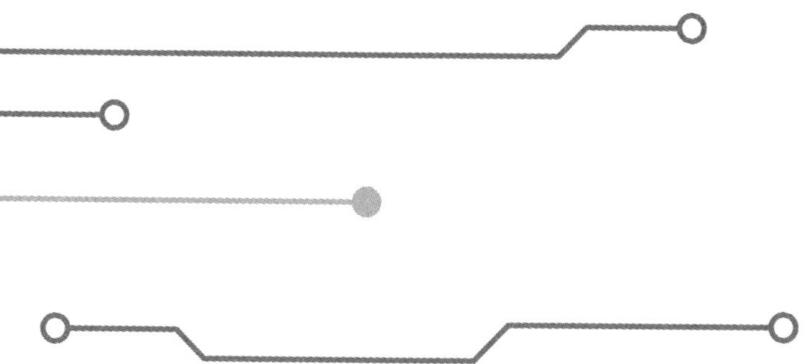

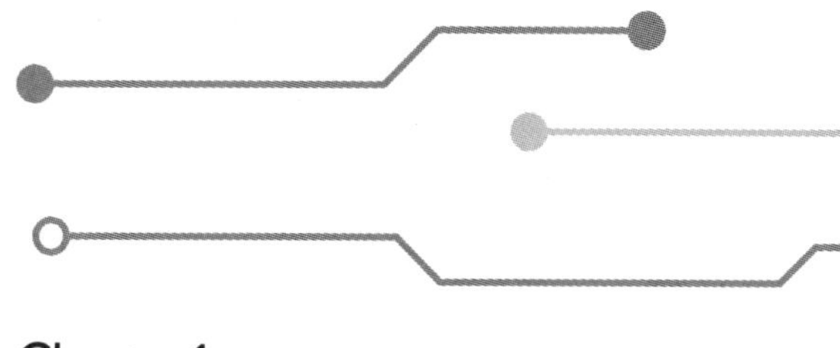

Chapter 1

CHAOS THEORY COMES TO RETAIL

In 1952, Ray Bradbury, the famous science-fiction author, published a short story titled, "A Sound of Thunder." Set in the year 2055, time travel had become a real (and profitable) technology, and a group of affluent hunters goes back in time on a guided safari to kill a Tyrannosaurus rex. Although the safari company has strict procedures to minimize any changes that could occur in the future because of their actions—staying on the levitating path, only killing "tagged" species that were supposed to die within a few minutes anyway, not leaving any artifacts in the past like bullets—one hunter loses his nerve and steps off the path. Upon returning to the present, they find reality irrevocably changed: the English language has been altered, and the fascist candidate who originally lost has now won the presidential election. Looking at the mud caked under his boots, the hunter finds a single crushed butterfly.[*]

---

[*] Kaelen Encarnacion, "The Literal Butterfly Effect," *Nu Sci* magazine, September 20, 2021, https://nuscimagazine.com/the-literal-butterfly-effect/.

Chaos Theory

Knowledge of upcoming weather was of paramount importance to military planners during World War II, who needed to know if major storms would impede ground advances or bombing sorties. The military increasingly relied on meteorologists to aid them in their planning, and it is a little-known fact that perhaps one of the most important weather forecasts ever made was the one for D-Day, the day the Allies invaded France.

> *For the Allied invasion to have any chance of success, General Dwight D. Eisenhower, Supreme Commander of the Allied Forces, needed a full moon, a low tide, little cloud cover, light winds, and low seas. The low tide was necessary to allow soldiers to see, avoid, and disarm the mined obstacles. In June 1944, a full moon and low tide coincided on June 5, 6, and 7. The invasion of France had been scheduled for June 5, 1944.*
>
> *Outside, on the morning of June 4, the weather was quiet, with only a light breeze blowing. Despite this, the weather was unsettled, with a series of low-pressure systems developing out in the North Atlantic and heading toward the English Channel.*
>
> *One of Eisenhower's chief meteorologists for the invasion was Group Captain James Stagg. Stagg was only one of several meteorologists studying the weather for the invasion. Leading up to the mission, the team of meteorologists would confer regularly and disagree about the forecast. These meteorologists were operating without technology and equipment that today's forecasters take for granted, such as satellites, weather radar, computer models, and instant communications.*
>
> *On the afternoon of June 4, when the weather began to*

deteriorate as the first storm approached, Stagg noticed an observation from a single ship stationed 600 miles west of Ireland reporting a rise in the barometric pressure. The pressure kept rising. Considering this ship's report, Stagg deduced that there could be a break in the weather on June 6 that current and historic predictions were not accounting for.

Late on June 4, Captain Stagg met again with the Allied commanders while the storm raged outside. Stagg told Eisenhower that he thought the weather would improve and strongly suggested they push back the invasion to June 6. He shared there might be winds of force 3 or 4, rising to force 5 in a few places, but the sky should be clear by the time they invaded. He explained that it might cloud over later, but the cloud bases should stay high enough for the naval gunners to spot their shots. Not ideal, but good enough. Eisenhower concurred and pushed the invasion to June 6.

*That forecast was a pivotal moment in world history. If the forecast was wrong, the lives of thousands of men and massive amounts of equipment would be lost. But, on the other hand, if the unsettled weather forecast for June 5 had not manifested and the weather had been good, the Germans might have spied the massive build-up of forces along the coast of southern England. While far from perfect, the weather on the morning of June 6 was good enough for the invasion to proceed, and the rest, as they say, is history.**

---

* Adapted from "How a Weather Forecast Made History—the D-Day Landings," MetMatters: The Royal Meteorological Society, accessed September 2023, https://www.rmets.org/metmatters/how-weather-forecast-made-history-d-day-landings.

One of the young meteorologists working for the U.S. Army Air Corps during World War II was Edward Lorenz. Lorenz grew up in Connecticut and was fascinated by the weather. He recorded the temperature readings every day from the thermometer outside the kitchen window. He was also keenly interested in mathematics, going on to earn degrees from both MIT and Harvard in the subject.

During the war, Lorenz and his fellow forecasters were forced to use simple models of the weather built using manual calculations. The time-consuming, laborious process of working through manual calculations imposed strict limits on weather modeling. But that began to change in the 1950s when early computers rapidly evolved amidst the shift from vacuum tubes to transistors. That change meant that computers could become smaller, which quickly expanded their use beyond military weapon development into many other more accessible industries, including weather forecasting.

After the war, Lorenz returned to MIT to do graduate work in meteorology, combining his knowledge of mathematics with weather measurement and forecasting. Using the newfound computational power, Lorenz began to capitalize on his mathematical knowledge and meteorology expertise to begin building far more sophisticated weather forecasting models that became increasingly accurate and able to look beyond the next twenty-four to forty-eight hours.

By early 1961, Lorenz had built a model to simulate atmospheric convection, modeling the air currents that drive weather in any given area. After running the model, Lorenz wanted to redo one of the simulations. Entering the initial data again, he took a shortcut, only typing in three decimal places, like 0.506, rather than the six decimal places, like 0.506127, used in the initial run. The computer he was using only printed out variables to three decimal places but used six decimal points for the internal calculations.

Running the model a second time, he was shocked to find a completely different forecast. A tiny change in the initial conditions—shortening the inputs to thousandths from millionths—resulted in radically different system behavior. That insight led Lorenz to the development of what we know today as Chaos Theory.

> *Chaos Theory: the branch of mathematics that deals with complex systems whose behavior is highly sensitive to slight changes in conditions, so that small alterations can give rise to strikingly great consequences.*
>
> *—Oxford Language Dictionary*

Retail: Complexity Incarnate

Take a moment to consider and appreciate the vast fast-moving consumer goods retail industry; it is complexity incarnate. Millions of products, hundreds of thousands of stores, millions of workers, and supply chains stretching around the world. All of that is made even more complex by having to manage perishable fresh foods, sell-by dates on packaged products, product recalls, and more. Then consider managing all this complexity twenty-four hours a day, seven days a week, on impossibly thin profit margins. And most people think selling groceries is a pretty straightforward business.

In retail, we have a massive, dynamic, and intricate industry—a complex system in Lorenz's world. And we don't have to look far to understand the impact of a small change to that system. I'll never forget the surreal experience of walking into the Von's (Safeway) store in Marina del Rey, California, a short time after the COVID-19 pandemic shut down the country in March 2020. After standing in line outside to get into the store, we were directed via arrows on the floor to move through the store in a one-way flow to minimize interaction with the few other shoppers.

When I turned into the paper aisle, I was taken aback. Most of the shelves were empty, only a few rolls of paper towels, boxes of Kleenex, and paper plates were available. The toilet paper section had 16 feet of empty shelves. Turning the corner into the next aisle, where the canned goods like soup were located, I saw the same devastation.

As the shutdown continued, problems cascaded through the supply chain in a self-reinforcing cycle. Product shortages sparked panic buying. Panic buying screwed up demand forecasts, and retailers began ordering far more cases, knowing they might, at best, get only a fraction of that. Manufacturers were forced to pare back variety and brand extensions that had been built over decades due to supply and packaging shortages. Lack of personnel in distribution centers created more product shortages. Shoppers grabbed anything that was close to the products they were searching for, creating more empty shelves. Retailers attempted to limit purchases of items like toilet paper, beef, and chicken, but even that didn't seem to help.

Technology woes exacerbated the problems. Spiking demand for certain products, like yeast (as millions of people began baking at home almost overnight), while other products stopped selling, overwhelmed retailers' computer-assisted ordering systems. Out-of-stocks not being reflected in real time in the retailer's eCommerce system made online shopping hell-on-earth for shoppers and for retailers seeking to pick and fulfill orders. Changing regulations, changing safety guidelines, and growing worker shortages as people became ill all added to the chaos caused by an invisible virus.

Doug Baker is vice president of industry relations for the Food Industry Association (FMI), and he relates a story highlighting the impact of a seemingly tiny disruption on the massive food industry.

It turns out nails are vital for our food.

As the pandemic entered full swing amidst government-ordered shutdowns, professional remodeling and renovation projects came to a grinding halt, but at the same time, DIY projects exploded as people sought to improve their living spaces themselves. One of the results of the DIY renovation surge was that nails quickly became in short supply. Domestic nail manufacturers were overwhelmed by the demand and simply could not keep up. Making matters worse, foreign nail producers couldn't get their goods into the U.S. market due to the massive backlogs at the shipping ports.

So, what do nails have to do with the food we eat? Nails are used to make pallets, those wood platforms that can be stacked with cases of goods and then moved by forklifts or pallet jacks. Nearly all the food products you find in a grocery store are transported from their place of production or manufacture to distribution centers and then on to the store, on pallets. It turns out that pallets are vital to the distribution of food.

The nail shortage meant that pallets could not be built or repaired. Making the situation even more challenging, retailers increased the use of pallets for displays in the store, putting pallets of paper towels or toilet paper on the sales floor for shoppers to pick up. So fewer pallets were being built and repaired, and more pallets were being tied up as part of displays at retail.

Lack of pallets meant a growing number of cases of product piling up in the manufacturer's distribution centers. As those warehouses filled up, production lines were slowed down and sometimes even halted. There were empty shelves across tens of thousands of stores, all for the want of nails. This is chaos theory in action. A complex system was highly sensitive to a seemingly insignificant shift in availability, giving way to massive consequences that rippled across the world.

The pallet shortage is a great example of the disruption caused by the COVID-19 virus, and it also exposed the complexity and frailty of producing, packaging, and distributing products. Shortages of certain beverages and foodstuffs occurred not because of ingredient shortages but because aluminum cans came into short supply. Pickles were in short supply at both the grocery store and fast-food restaurants because glass jars became unavailable. A glue shortage meant that product manufacturers could not seal boxes to ship their products.

Another of these sought-after products was flour for baking. King Arthur Baking Company, known for its eponymous baking flours, is a more than 230-year-old company founded in 1790. Think about that history: the company was founded when there were only thirteen states in the United States. The company's flour was used in sourdough bread, feeding the miners during the gold rush in 1849. King Arthur Baking Company has endured the American Civil War, two world wars, and many other conflicts. The company not only survived but thrived throughout its tumultuous history. And yet King Arthur Baking Company was nearly shut down by an invisible foe: COVID-19.

I spoke with co-CEO Karen Colberg in early 2021, nearly a year into the pandemic, about what the company had gone through and how they had survived. From spiking demand as home-bound people began baking, to the shutdown of many of the retail businesses they supplied, the experience was one of extremes. Like other manufacturers, King Arthur struggled to obtain supplies—in their case, wheat for milling—and packaging materials. They struggled to maintain their workforce; office people could work remotely but milling required hands-on work. King Arthur Baking Company's experience reinforced just how delicate the supply chain was for a seemingly straightforward business.

The pandemic drove home to every consumer—all 330 million of us in the United States—just how fragile the consumer goods supply chain is. For the first time in many people's lifetimes, the goods and products they had taken for granted were no longer readily available. The pandemic exposed not just how fragile the grocery supply chain was (and still is), but the global scale and the intricacies of supplying millions of products to hundreds of millions of consumers. COVID-19 showed us how the slightest hiccup in this massively complex and fast-moving industry can wreck havoc . . . almost like a hurricane coming through.

The Butterfly Effect

Lorenz's work wasn't finished with the epiphany leading to Chaos Theory. Digging deeper, he graphed the data he collected in his weather-prediction models, noting some strange properties that began to appear. Saving readers the complexity of the mathematics,

Lorenz found in graphing the data that the dynamics of his equations served two purposes: repulsion of trajectories within the data set and then attraction beyond it.* The result of graphing the data was a chart that looked like that below:†

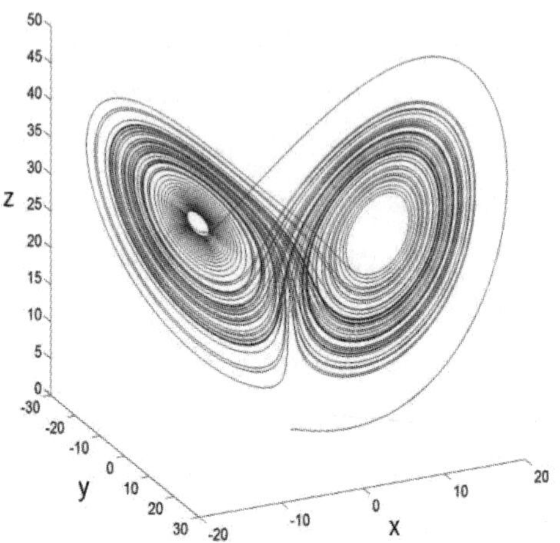

Does that shape look familiar? Combining the discovery of minute changes leading to dramatic and unpredictable results and the appearance of the charted data, Lorenz coined the term "butterfly effect." The butterfly effect effectively states that if a butterfly flaps its wings in one part of the world, it will ultimately trigger a

* Paul Halpern, "Chaos Theory, the Butterfly Effect, and the Computer Glitch That Started It All," *Forbes*, February 13, 2018, https://www.forbes.com/sites/startswithabang/2018/02/13/chaos-theory-the-butterfly-effect-and-the-computer-glitch-that-started-it-all/?sh=6e2d58c869f6.

† Samuel Toluwalope Ogunjo, "Increased and Reduced Order Synchronization of 2D and 3D Dynamical Systems," *International Journal of Nonlinear Science* 16, no. 2 (2013): 105–112.

tornado somewhere else.* Not only do small changes in nature have unpredictable effects, but it is also impossible to predict the resulting impact.

The idea was cemented into mainstream culture in a scene from *Jurassic Park*, where Jeff Goldblum's character explains chaos theory. "It simply deals with unpredictability in complex systems," he says. "The shorthand is 'the butterfly effect.' A butterfly can flap its wings in Peking, and in Central Park, you get rain instead of sunshine."†

The Butterfly Effect Comes to Retail

Imagine that each of the hundreds and thousands of new innovative technologies, solutions, and capabilities flooding into the retail industry right now is a butterfly. And any one of those butterflies has the potential to trigger hurricane-strength disruption in your business. " The time between the emergence of a butterfly and the ultimate disruption grows shorter every day, giving you less and less time to adjust.

Generative AI is an innovation butterfly, and its impact is growing by the day as more and more companies find ways to use this capability across their organizations."

---

* Nethra Iyer, "A Method to Madness: The Mathematical Basis for the Butterfly Effect," December 6, 2020, https://nuscimagazine.com/a-method-to-madness-the-mathematical-basis-for-the-butterfly-effect/.

† Jeremy Deaton, "The Butterfly Effect Is Not What You Think It Is," *Washington Post*, February 2, 2020, https://www.washingtonpost.com/weather/2020/02/02/butterfly-effect-is-not-what-you-think-it-is/.

Solutions like ChatGPT, perhaps the most well-known example of generative AI, are transforming search, marketing, content creation, and more, as companies like Instacart integrate generative AI capabilities to help shoppers with meal planning and recipe discovery.

And there are more new butterflies emerging every day. This is precisely the world that the massive retail industry has now entered. A world where the risk of disruption grows exponentially in lockstep with technology.

Scott Brinker is the founder and editor of *Chief MarTech*, a blog devoted to marketing technology. Some years ago, he developed an annual marketing technology landscape infographic to identify all the solution providers in the space. Looking at how marketing technology has grown over the years showcases just how fast innovation is happening. In 2011, the marketing tech landscape consisted of 150 solution providers. By 2022, there were 9,932 solution providers, representing a 6,521 percent growth over the past ten years.*

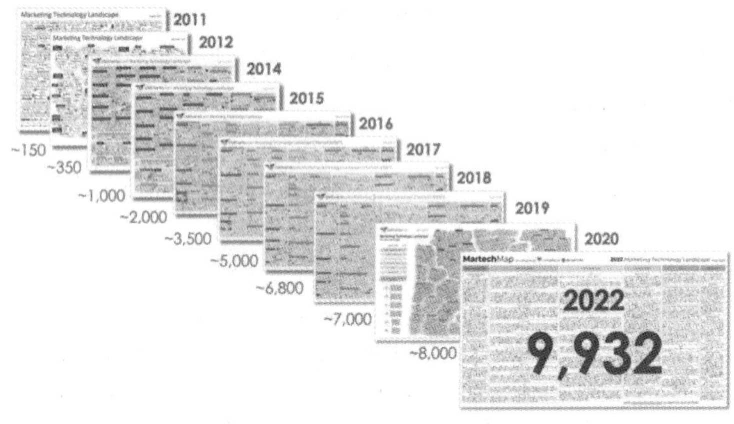

* Scott Brinker, "Marketing Technology Landscape 2022," Chief MarTec, accessed September 2023, https://chiefmartec.com/2022/05/marketing-technology-landscape-2022-search-9932-solutions-on-martechmap-com/.

And that's just in *marketing*. Think about all the other innovation we've seen in the retail industry over the past decade. Automated distribution centers, stores without cashiers, shoppers with super computers in their hands, automated business processes, the list goes on and on.

And we're just getting started.

Chapter Summary

- Retail is a massively complex industry, and seemingly minute changes—such as a shortage of the common nail—can release a tsunami of change across the entire industry without notice. As tech innovation accelerates, the probability of any given new innovation to cause disruption increases.
- Disruption in a complex system is inherently unpredictable. As such, companies would be well-served to adapt their organizations and cultures to this new world, to be better able to absorb unexpected change when it inevitably happens.
- Becoming nimble and creating an 'early warning system' to identify new disruptive technologies and capabilities as early as possible, can help position your organization for the chaos of new innovation.

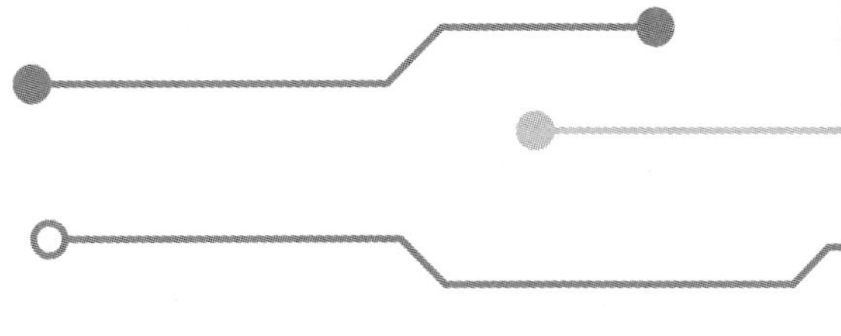

Chapter 2

THE INNOVATION FLYWHEEL

Every day you read about new innovations, and many times these reports are about things you never thought were possible. Just today, I read that Elon Musk's humanoid robot, Optimus, can now walk, pick up, and carry items and perform basic tasks. Musk is predicting that one day there may be more robots than humans.* Another company has developed an electronic "skin" that can allow robots to sense touch and their surroundings. The science of longevity is advancing by the day with new discoveries around supplements and therapies to extend productive lifespans to 100, 120, or even more years. Amazon's Zoox, the company's autonomous car division, is now taking its fleet of robotaxis on public roads. Human brain cells are being used to create a new kind of computer. Another

---

* Scarlett Evans, "Tesla's Humanoid Robot Can Now Walk; Musk Predicts They'll Outnumber Humans," IOT World Today, March 3, 2023, https://www.iotworldtoday.com/robotics/tesla-s-humanoid-robot-can-now-walk-as-musk-predicts-they-ll-outnumber-humans.

company has developed the ability to 3D print living cells inside the body, which will hopefully someday allow for doctors to replace damaged cells and diseased organs within a living human body. And that's just today. Imagine what tomorrow may bring.

Peter Diamandis, the founder and executive chairman of the XPRIZE Foundation, and executive founder of Singularity University, puts it this way: "This enormous growth in technology will entirely reconstruct the business world as we see it today. Enterprises that take advantage of technologies such as artificial intelligence, 3D printing, blockchain, quantum computing, nanotechnology, AR/VR, robotics, etc., will have the most dynamic growth in this decade. As these technologies converge and grow together, businesses that implement these tools will become more efficient and save a lot of capital that could be invested elsewhere."*

Do you ever wonder why we're seeing more and more innovation everywhere we look? What's behind this? And why now? I believe understanding the "why" of innovation can help retail leaders make sense of the "what." Understanding can help assimilate the new information arriving every day in our fast-changing world.

There are three fundamental, foundational, forces—each growing exponentially—that are powering up the transformation and disruption we're seeing in all parts of our lives. These three things are:

---

* Andrew Kirima, "How the Acceleration of Technology Will Impact the Business World," Medium, August 18, 2020, https://medium.com/nerd-for-tech/how-the-acceleration-of-technology-will-impact-the-business-world-976b72122921.

- ever-faster and ever-cheaper computer processing power;
- ever-expanding data; and
- ever-more powerful artificial intelligence

These three components come together to create what I think of as the *innovation flywheel*, a self-reinforcing, ever-expanding cycle of new capabilities having an ever-greater impact on our world.

Computer Processing Power

There is perhaps no one better to explain technological advancement than Ray Kurzweil. Kurzweil is a renowned author, noted inventor, and futurist who leads Google's AI initiatives around the automated understanding of human language. "Of the 147 predictions that Kurzweil has made since the 1990s, fully 115 of them have turned out to be correct, and another 12 have turned out to be 'essentially correct' (off by a year or two), giving his predictions a stunning 86 percent accuracy rate."* Given that success rate, Ray Kurzweil is someone to listen to when it comes to technology.

Kurzweil has plotted the growth of computer processing power over the past century. To make this growth easier to understand, Kurzweil equates the processing power at different points in time to that of a $1,000 computer. In 2001, the processing power of a $1,000 computer was equivalent to the brain of one insect. Fifteen

---

* Peter H. Diamandis, "Ray Kurzweil's Wildest Prediction: Nanobots Will Plug Put Brains into the Web by the 2030s," Singularity Hub, October 12, 2015, https://singularityhub.com/2015/10/12/ray-kurzweils-wildest-prediction-nanobots-will-plug-our-brains-into-the-web-by-the-2030s/.

years later (2015), it was comparable to the brain of a mouse. Today, a $1,000 computer has the processing power equivalent to a human brain. Within the next twenty or so years, that same computer will have processing power equivalent to the combined brains of every human being on planet Earth—a projected 9 billion people.

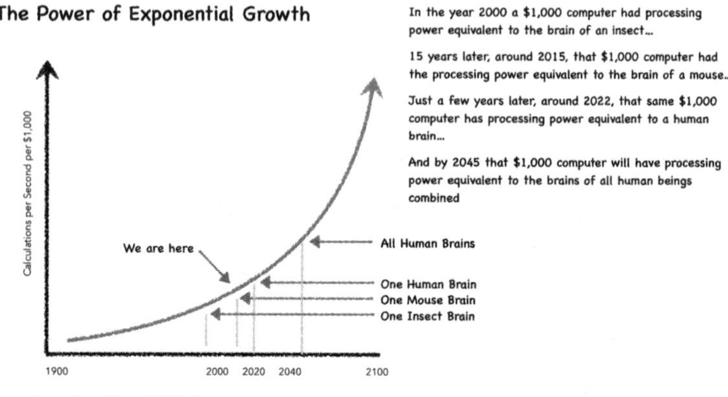

Stop to think about that. Approximately twenty years ago, that $1,000 computer had the processing power of an insect. But twenty years from now, that $1,000 computer is projected to have processing power equivalent to 9 billion human brains. Think about all the change we've experienced in the past twenty years: the smartphone and app stores, self-driving cars, the growth of the World Wide Web, new businesses like Instacart and Uber, and far more. And that's all happened because of the growth of computer processing power.

Growth of Computer Processing Power Is Accelerating

On October 14, 2012, Felix Baumgartner opened the door of his cabin, took a look around, and stepped out, plummeting to the earth nearly 24 miles below. Baumgartner was sponsored by Red Bull to complete a "space jump" as part of Red Bull's Stratos project. He jumped out of a helium balloon from the stratosphere and traveled four minutes and twenty seconds to Earth's surface, landing in New Mexico. In the first 20 seconds, he accelerated to a velocity of 700 km/hr, and over the next several minutes, he continued to accelerate to a maximum speed of 834 mph, breaking the sound barrier along the way, until the denser air began to slow his descent.[*]

Unless you've gone skydiving, the idea of continuous acceleration is foreign to most people. As human beings, we quickly reach the limits of continuous acceleration, be it running as fast as we can from a standing start or driving a Formula 1 race car off the starting line. The physical world has limitations. At least it used to.

---

[*] Phil Plait, "Fall 40 Kilometers to Earth in a POV Video from Felix Baumgartner's Helmetcam," Slate, October 15, 2013, https://slate.com/technology/2013/10/stratospheric-jump-helmetcam-video-of-felix-baumgartners-jump.html.

In the world of technology, computer processing power has been accelerating since the creation of the first mechanical tabulating machines. The pace really picked up in 1958 when Bell Labs announced the first transistor, which launched the electronic age. Transistors are miniature semiconductors used to control the flow of electronic signals. Before their invention, electronics were dependent on vacuum tubes, which were much larger and unreliable. The invention of transistors allowed for faster, more reliable, and more powerful electronics. Not long after, in 1965, Gordon Moore observed that the number of transistors in an integrated circuit was doubling about every eighteen months, giving rise to Moore's Law, which many readers will be familiar with. It was this exponential growth that gave rise to the semiconductor industry and has powered the growth of technology for the past nearly sixty years.[*,†]

[*] "What Is Moore's Law?" Synopsys, accessed September 2023, https://www.synopsys.com/glossary/what-is-moores-law.html.

[†] The Beauty and Joy of Computing, "History and Impact of Computers: Moore's Law," accessed September 2023, https://bjc.edc.org/March2019/bjc-r/cur/programming/6-computers/3-history-impact/2-moore.html?topic=nyc_bjc%2F6-how-computers-work.topic&course=bjc4nyc.html&novideo&noassignment.

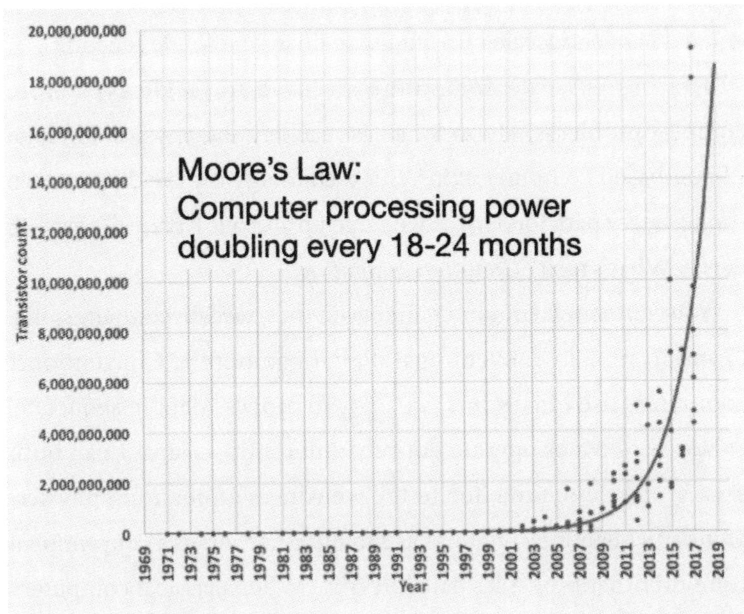

Think about this: The first transistor-based computer chips had only a handful of transistors. Apple's M1 chip (they're now working on the M3 chip) has *40 billion* transistors on it.

Exascale supercomputers represent the next, immediate leap in computing. Exascale computers are digital computers, just like the desktop and notebook computers we use today, but with much more powerful hardware. To give you a sense of the scale of these machines, consider this: A notebook computer like the one you probably use is capable of processing around 1 trillion operations per second. Sounds impressive, right? But the Oak Ridge National Laboratory's Frontier supercomputer can process more than 1,000,000,000,000,000,000 operations per second. "To contextualize how powerful an exascale computer is, an individual would have to perform one sum every second for 31,688,765,000 years to equal what an exascale computer can do in one single second."*

Even as exascale computers are used to tackle ever more complex problems, Moore's Law is reaching the physical limits of silicon-based computer chips. But technology is providing its own evolutionary path forward as we surge up and out Kurzweil's growth curve. What's next? *Quantum computing.*

Your notebook computer, and even the exascale computers like Frontier, remain conventional digital computers. Conventional computers use binary bits, a 0 or 1, to process data in sequence. Quantum devices operate entirely differently—0 and 1 can both exist at the same time due to the weirdness of quantum physics. Think of a quantum computer as a multitasker versus a conventional computer that processes data linearly. "When classical computers solve a problem with multiple variables, they have to conduct new calculations every time a variable changes. Each calculation is a single path to a single result. On the other hand, quantum computers have a larger working space, which means they can explore a massive number of paths simultaneously. This possibility means that quantum computers can be much, much faster than classical computers."†

Quantum computing's ability to simultaneously process multiple bits of information makes it tailor-made for the complex world of retail. Use cases have already been envisioned as the technology continues to mature include analysis of customer purchases,

---

* McKinsey & Company, "What Is Exascale Computing?" November 22, 2022, https://www.mckinsey.com/featured-insights/mckinsey-explainers/what-is-exascale-computing.

† Ibid.

the routing of delivery trucks flowing from manufacturing plants to distribution centers and on to stores, demand forecasting, labor scheduling, and more.

Quantum computing is already coming to retail. Save-On Foods, a Canadian grocer, is working with quantum computing company D-Wave to solve a logistics problem with 4 million decision variables that overwhelmed conventional computers. D-Wave developed a hybrid quantum algorithm that is already running in stores, reducing the processing time for some tasks from twenty-five hours a week to just seconds. Initial work has been so promising that the retailer is expanding the technology to other stores and other operational initiatives.*

Burgeoning Big Data

In 2020, I released the white paper *Retail 4.0: The Age of Metamorphosis,* which spoke to the digitalization of the massive retail industry.† Digitalization equals data, and it's not just the volume of data being produced, it is the exploding diversity of the data that is transforming everything from manufacturing and production to

---

* Daphne Leprince-Ringuet, "Quantum Computers Are Coming. Get Ready for Them to Change Everything," ZDNet, November 2, 2020, https://www.zdnet.com/article/quantum-computers-are-coming-get-ready-for-them-to-change-everything/.

† Gary Hawkins, "Retail 4.0: The Age of Metamorphosis—A Briefing for Retail Industry Executives on a Future Arriving Sooner than Expected," Center for Advancing Retail & Technology, 2020, https://www.advancingretail.org/retail-40-paper-the-age-of-metamorphosis.

distribution, operations, and marketing.

Retail's use of data began to accelerate in the 1980s, the barcode enabled the tracking of individual SKUs, helping give rise to product category management. Barcodes on individual products and product cases helped Walmart's legendary logistics expertise. The barcode spread to shoppers via frequent shopper cards, enabling the retailer to effectively scan shoppers and building a repository of shopper-identified purchase history data.

As digital capabilities grew, online behavior data was added to the mix. The smartphone has given rise to location data providing insight to shopper behavior across the marketplace. Growing product data, especially health and nutrition related attributes, has given rise to even greater shopper insights and more impactful marketing personalization.

So, as you can see, data has always fueled the retail industry, but its role in the innovation flywheel has become increasingly important as data feeds new, increasingly AI-powered capabilities. And those new capabilities increasingly help drive revenue gains, cost decreases, and monetization opportunities.

Look what Amazon is doing. Using the data provided by its Just Walk Out and Dash Cart (smart cart) technologies, Amazon has created an analytics platform where suppliers can access anonymized and aggregated insights across Amazon Fresh stores about how shoppers engage with the store, categories, specific products, and product purchases. This kind of data powers more effective shelf sets, the placement of products within a category, and informs package design (which label design impacts sales the most).

As I've written about before, computer vision platforms like Just

Walk Out and smart carts provide an entirely new stream of data to retailers and their vendors about how shoppers act within the brick-and-mortar store. Linking new scorecards of in-store behavioral data to merchandising opens the door to measuring the impact of signage, special displays, ad features, and more. One study we conducted showed significant increases in aisle traffic and sale of adjacent products if specific brands of cereal were promoted. And, in the case of Amazon, this data creates new monetization opportunities as CPGs and vendors use the store analytics platform.

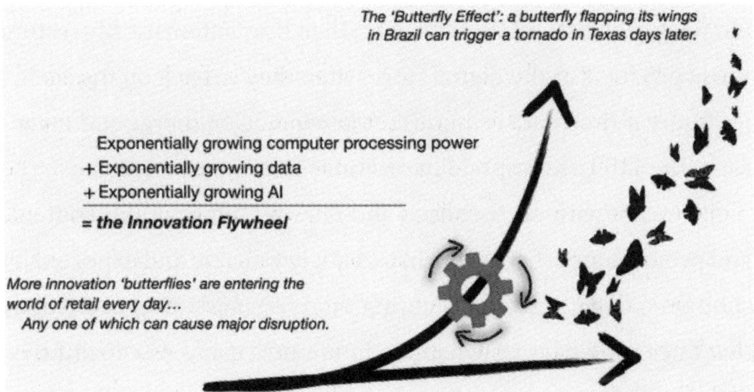

Retail media networks (RMN) are an increasingly important source of revenue for retailers, particularly as they look to offset costs incurred by eCommerce operations. And it is data that drives RMN revenue, not just the ability to present ads to shoppers while they're shopping online or using the retailer's app. Kroger, for example, has built out a massive ad platform that enables advertisers to build and target audiences using anonymized and aggregated shopper purchase data; Kroger tracks purchases on over 60 million households. The company then links those audiences to myriad vehicles to communicate with the shopper. Then they close the loop with full attribution, letting the advertiser see how many shoppers saw an ad, visited the store, and then made a purchase. More data.

Data is also becoming more real-time, something that's especially important relative to inventory. As retailers grapple with making eCommerce profitable, searching for out-of-stock products and then dealing with substitutions represent a significant cost. Leading retailers are working to link real-time inventory systems to their digital storefronts, automatically removing products that are out-of-stock to eliminate downstream inefficiency. Then they automatically restore those products to the digital store when they're back on the shelf.

But real-time data is important beyond eCommerce and inventory. The ability to respond in real time to shoppers engaging in the digital realm with personalized and relevant information, content, and promotions is becoming increasingly valuable and expected by shoppers. Creating and executing targeted marketing campaigns that once took days now happen in minutes using AI capabilities, and they deliver greater efficacy. And marketing personalization is increasingly being extended to pricing, with retailers providing different discounts on a given product to different shoppers based on what they know about the customer. In this world, retailers need the ability to manage promotions, pricing, and quantity limits, across channels in real time and all at scale.

And, this being retail, we have to address the cleanliness and accuracy of the data. Fast-moving consumer goods retailers have a long history of poor data discipline, resulting in part, unbelievably, from the product descriptions used in the early POS scanning systems with their character limitations. But that poor discipline grew into poor maintenance relative to product categorizations and other attributes. While there has been some good progress made, much work needs to be done.

Data cleanliness is a massive challenge for even the largest retailers who throw millions of dollars a year at the problem. Product data and related attributes, especially health and nutrition attributes, are becoming increasingly important as the food-as-medicine trend grows. Sifter Solutions is working with some of the largest retailers to clean and manage their data, using Sifter's advanced AI and machine learning capabilities. Clean data powers the world of Bionic Retail.™

Retail executives should view data as a strategic asset, not just something collected as a byproduct of other systems. Data, especially shopper intelligence and creating a digital twin of the store and all the products within it, is critical as digital engagement and other technologies like virtual reality and augmented reality grow.

Retail leaders need to ensure they are structuring data correctly. "According to a global survey (by Salesforce) last year, 33 percent of business leaders said they can't generate meaningful insights from their data, and 30 percent said they were overwhelmed by the sheer volume."* The *Wall Street Journal* article goes on to state that "the stakes are higher than ever now, tech leaders said. Companies that don't have a strong grip on their data will struggle to customize and leverage the possibilities of artificial

* Isabelle Bousquette, "Data Deluge: Businesses Struggle with TMI," *Wall Street Journal*, March 20, 2023, https://www.wsj.com/articles/data-deluge-businesses-struggle-with-tmi-5e41cca1.

intelligence—including rapidly developing generative AI models, like the newly released GPT."*

Artificial Intelligence

> *"Companies have to race to build AI, or they will be made uncompetitive. Essentially, if your competitor is racing to build AI, they will crush you."*
>
> —*Elon Musk*

Artificial intelligence (AI) has long lurked at the periphery of business, a technology of great promise but never quite ready for prime time—until now. AI systems are powered by ever-growing processing power and feeding on ever-more data, and they are bursting forth, transforming existing capabilities and creating never-before possible abilities. Indeed, Peter Diamandis claims that AI will achieve human levels of intelligence by 2029. Elon Musk is even more aggressive; he feels that AI will be vastly smarter than any human and can overtake us by 2025.†

Within ten days of ChatGPT's launch, it had one million users; within sixty days, more than 100 million people were using it. ChatGPT has brought artificial intelligence into the boardroom, as millions of executives have experienced for themselves the power of AI. And that

---

* Ibid.

† Singularity University, "Abundance 360: Top 20 Metatrends & Moonshots for 2022-2032," accessed September 2023, https://www.diamandis.com/hubfs/METATRENDS%20%26%20MOONSHOTS%20-%20V2.pdf?.

power can be frightening. The first time I "conversed" with ChatGPT, it truly felt like there was an intelligent being on the other side of the keyboard. As I would ask a question, the AI would provide the response. I'd ask another question, and so on. It was like Star Trek's Computer had come to life. And that conversation has moved from the keyboard to voice interaction as ChatGPT is voice-enabled, powering ongoing conversations with the generative AI.

As millions of retail executives begin to understand the power of generative AI, we are seeing this technology explode into the marketplace. Every day, people are finding new ways to use this kind of AI to do a growing number of jobs. "ChatGPT could hypothetically pass interviews to be hired as an entry-level software engineer at Google, CNBC reported . . . citing internal Google tests of the tech."* A user asked ChatGPT what jobs it could replace. The system responded with a list including data entry clerks, customer service representatives, proofreaders, paralegals, bookkeepers, translators, and more.

> *"Artificial intelligence could have more profound implications for humanity than electricity or fire."*
>
> —*Sundar Pichai, CEO, Alphabet*

Considering the power of generative AI like ChatGPT, a growing number of solutions are coming into the market using

---

* Sawdah Bhaimiya, "ChatGPT Could Hypothetically Get Hired as an Entry Level Coder If It Interviewed at Google, Internal Document Reportedly Says," February 1, 2023, https://www.businessinsider.com/chatgpt-google-could-get-hired-as-coder-internal-document-report-2023-2.

domain-specific data and knowledge to fine-tune capabilities for different business functions. Using a company's own data also helps ensure that the responses provided by the AI are relevant and reliable. Imagine creating a generative AI solution that mimics a category manager's role, an AI capable of making decisions around what new products to carry, which products to delist, how much inventory to stock, and more.

So just what is "artificial intelligence"? Many people use the term but don't truly understand just what AI is. To answer that question, I went to the source: ChatGPT itself.

Hawkins: What is AI? How would you define it?

ChatGPT: Artificial Intelligence (AI) is a branch of computer science that aims to create intelligent machines that can perform tasks that typically require human-like intelligence, such as visual perception, speech recognition, decision-making, and language translation. AI systems are designed to simulate human cognitive processes, such as learning, reasoning, and problem-solving, and can be used to automate complex tasks that would normally require human intervention.

AI systems are typically built using algorithms and models that are trained on large data sets, allowing the system to learn from experience and improve its performance over time. There are different types of AI systems, ranging from rule-based systems that follow a set of predefined rules to deep learning systems that use neural networks to learn from data.

AI is a rapidly evolving field that has the potential to revolutionize many industries, from healthcare to finance to transportation. As AI technology continues to evolve, it will become increasingly important for businesses and organizations to understand its capabilities and limitations, and to find ways to leverage this technology to improve their operations and services.

Hawkins: Why is artificial intelligence so powerful?

ChatGPT: Artificial intelligence (AI) is powerful for several reasons:

- Speed and Efficiency: AI can process vast amounts of data quickly and efficiently, making it ideal for handling complex tasks that would take humans much longer to complete. AI systems can analyze data and make decisions in real-time, which can improve the efficiency of many industries.
- Accuracy and Precision: AI algorithms are designed to be highly accurate and precise, which makes them useful for tasks that require high levels of accuracy, such as medical diagnosis, weather forecasting, and financial modeling.
- Adaptability: AI systems can adapt to new situations and learn from their experiences, making them versatile and adaptable to changing circumstances. This adaptability makes AI valuable for tasks that require

flexibility and the ability to learn from new data.

- Consistency: AI systems can perform tasks consistently without getting tired or making errors due to fatigue or boredom. This consistency makes AI valuable for tasks that require a high degree of precision and accuracy.
- Scalability: AI systems can be scaled up or down depending on the size of the data sets and the complexity of the task at hand. This scalability makes AI valuable for large-scale projects that require the processing of massive amounts of data.

Overall, AI is powerful because it can perform complex tasks quickly and accurately, adapt to new situations, and scale up or down to meet the needs of different projects. These capabilities make AI a valuable tool for many industries and have the potential to revolutionize the way we live and work.

"I summarize it like this: There will be two kinds of companies at the end of this decade . . . Those that are fully utilizing AI, and those that are out of business."

—Peter Diamandis

Converging Technologies: The Law of Accelerating Returns

The Large Hadron Collider (LHC) is the world's largest and most powerful particle accelerator. The "accelerator propels charged particles, such as protons or electrons, at high speeds, close to the speed of light. They are then smashed either onto a target or against other particles circulating in the opposite direction. By studying these collisions, physicists are able to probe the world of the infinitely small. When the particles are sufficiently energetic, a phenomenon that defies the imagination happens: the energy of the collision is transformed into matter in the form of new particles, the most massive of which existed in the early Universe."*

The story of retail's future is not only being written by accelerating technologies. Like particles in the LHC, exponentially growing technologies are crashing together and giving birth to new mind-blowing capabilities. Quantum computers are a result of this kind of convergence, powered by increasing knowledge of quantum physics, advances in materials science, and software programming.

We're seeing the dramatic advancements in computer vision come to life. For example, Amazon is expanding its Just Walk Out technology into a growing number and size of stores. Ever-increasing computer vision sophistication is driven by the convergence of faster, cheaper processing power, more accurate and cheaper camera sensors, ever-faster networking speeds, the growing sophistication of artificial intelligence, and more.

Ray Kurzweil has studied this convergence and how colliding technologies accelerate technological growth even further. According to *AI Time Journal*, "the Law of Accelerating Returns is an evolutionary process—this means that a) the rate of technological innovation is exponential, and b) as technological progress is made

---

* CERN, "Accelerators," accessed September 2023, https://home.cern/science/accelerators.

throughout shorter and shorter timespans, the use of new technologies to create novel ones accelerates the rate of exponential growth itself. This means there are two positive feedback loops at work here: one that dictates the initial rate of innovation and another that dictates the rate of innovation, both of which are exponential."*

The convergence of exponentially accelerating technologies is in turn driving the convergence of previously disparate industries. The growing convergence of the massive food and healthcare industries is a good example. Cutting-edge nutrition science, exploding data, and AI are powering new abilities to guide individuals to specific food products beneficial to their health. For example, a shopper—let's call her Haviland—can build a profile in a retailer's app indicating she has gluten and dairy allergies. Those attributes then filter the retailer's products, enabling Haviland to quickly find appropriate products in each product category when she's shopping online or in the store's aisle.

This food-as-medicine approach is transforming the healthcare and health-insurance industries as everyone searches for a way to reign in massive healthcare costs while improving the well-being of people. This isn't just allowing some people to live longer; this advancement is using a deep, scientific approach to food guidance for each individual to improve the quality of their lives—people are

---

* Sasha Cadariu, "The Law of Accelerating Returns, Superintelligence, and the Technological Singularity, *AI Time Journal*, October 6, 2022, https://www.aitimejournal.com/the-law-of-accelerating-returns-superintelligence-and-the-technological-singularity#:~:text=The%20Law%20of%20Accelerating%20Returns%20is%20an%20evolutionary%20process%20-%20this,rate%20of%20exponential%20growth%20itself.

living better more fulfilled lives for longer. This is a major innovation affecting the human experience.

In short, what this means is that technological advances feed on themselves, further accelerating them. Exponential growth on top of exponential growth equals more butterflies. Think about that tomorrow morning when you have your first cup of coffee.

Chapter Summary

- Three fundamental forces are powering transformation and disruption: ever-faster and ever-cheaper computer processing power, ever-expanding data, and ever-powerful AI. The three of which make up the **innovation flywheel**.
- Retail executives must foster and protect a respect and understanding of their organization's data to ensure they are poised to make the best decisions in the wake of constant change in the exponential world.
- Although the myriad innovations surfacing may be dizzying, it's important to realize that the law of accelerating returns is at play, and each of those innovations will fuel the next generation of technology that much faster.

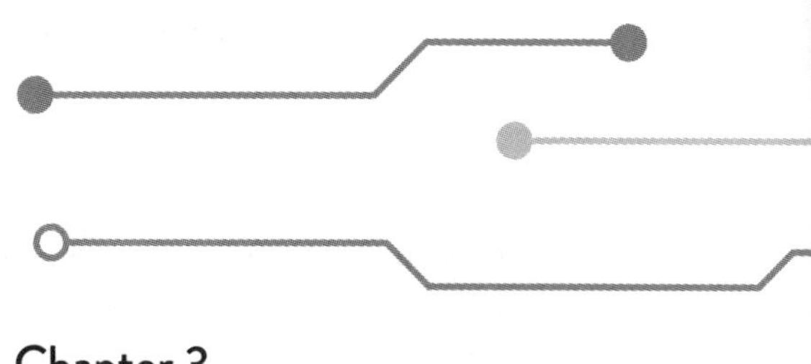

Chapter 3

TECH GESTATION GETTING FASTER

About ten months ago, we got a Bernese mountain dog puppy. We had several Berners when we lived back East when our kids were growing up. Now that we live in Colorado, it seemed only appropriate to get another. We got Remington when he was eight weeks old and weighed 10 pounds. As I'm writing this, he just turned one year old, and in the ten months we've had him, he's gained more than 100 pounds. So, Remington's size just exploded—that's over 1,000 percent growth in ten months. And the scary part is that he's not done growing!

Thinking about Remington's fast growth got me thinking about the growth of various technologies. And as I started doing some research, it became apparent that 1) in general, technologies are growing faster, hitting significant scale faster than in the past, and 2) the gestation period of new technology—the time from the initial idea to actual implementation—is also growing faster than it has historically.

Remington as a two-month-old puppy (left) and Remington at twelve months (right)

So why is this important? The innovation flywheel is spinning faster every day, producing more and more innovation butterflies. So not only do retailers have to try and keep track of a rapidly growing number of new innovative capabilities, but the time from the initial idea, to creating, implementing, and scaling new capabilities is growing faster, giving retailers less and less time to discover, understand, test, and deploy new capabilities. In the post-inflection-point world, speed doesn't kill—being slow does.

The UPC Barcode

The UPC barcode is perhaps one of the most significant and impactful technologies developed and deployed across the retail industry in the past century. Its development and widespread usage changed how retailers track inventory and operate internally. The development path of the barcode, from genesis to initial implementation to scaling, is an interesting story and a good example of how innovation often develops.

The History of the Barcode

The story of the UPC barcode begins in the 1940s, triggered by a "distraught supermarket manager who had pleaded with a dean at Drexel Institute of Technology in Philadelphia to come up with some way of getting shoppers through his store more quickly."

Ultimately, Joe Woodland, an inventor, heard of the challenge and, while sitting on the beach pondering the dilemma, idly ran his fingers through the sand, making four lines - the epiphany ultimately giving rise to today's barcode. He and a partner went on to file a patent in 1949 around the basic concept.

But the barcode itself was just the beginning. It took the development of the laser scanner to read the barcode to advance the idea. And it was Hughes Aircraft Company that built the first laser and announced it in 1960, little knowing it would help revolutionize the massive grocery industry.

Then in 1966, the Kroger Company, already one of the largest supermarket operators in the U.S., began a search for a way to make the checkout process faster and more efficient, intent on finding an answer to the decades-long dilemma. And it was IBM who ultimately brought all the pieces together; the barcode as we know it today, a laser scanner to `read' the barcode, and a mini computer to link the UPC number to an item description and price that would appear on the shopper's receipt.

The story of the UPC barcode is one of innovation and technological convergence. Woodland's patented code was a novel idea but effectively worthless without a way to accurately and efficiently read the code when printed on products. It took the invention of the laser to make barcodes readable. His original idea had a longer gestation period than the technology we're seeing today, and it is world's away from the rate of adoption we're about to see in the marketplace. The barcode also required the minicomputer to translate the barcode to an item description and price, along with other product attributions like department and category. And above all, the innovation required a compelling business case driven by customer convenience, operational efficiency, and faster, more accurate inventory data.

I include this story to make this point: The massive impact of the barcode was not realized for decades. It took eleven years from the barcode patent filing to the invention of the laser. Minicomputers were nearly another decade after the laser. Even from the first scan of the Wrigley chewing gum package in 1974, it took years to expand the barcode to a large majority of packaged goods products. And it took even more time to extend barcode technology to case packs of product, pallets of product, truckload shipments, and shipping containers. Barcoded coupons soon followed. It was around 1990 when the first frequent shopper programs appeared (the forerunner of today's loyalty programs), giving the shopper a card or key tag with a barcode on it that identified the shopper each time they made a purchase.

This was a good example of how innovation in retail tech's early days progressed. But now a powerful new innovation happening today.

Computer Vision Platforms

In 2010, our family's supermarket—which had become a live lab to test and trial new capabilities—became one of the first stores to be wired for video analytics. This first-generation capability owed its creation to the early digital cameras, faster computer processing, and faster networking speeds necessary to move large volumes of data.

The system, which Retail Next (previously known as BVI Networks) provided, was able to digitize the actual store and the customers inside of it. We were able to measure the number of shoppers entering the store; the conversion rates to departments, aisles, and categories; and render heat maps of shopper activity. Very eye-opening stuff, especially when relating shopper traffic and dwell time to merchandising initiatives to truly understand the impact of displays, signage, and product placement.

The law of accelerating returns continues to power innovation. AI and machine learning were added to this technology to improve camera sensors and faster computer processing, which resulted in the computer vision platforms rolling out across retail today, technologies like Amazon's Just Walk Out capability.

Amazon understands technology's growth. In January 2018, Amazon opened the first Amazon Go store in Seattle. That first store offering Amazon's Just Walk Out technology was around 1,800 square feet in size and offered a mix of packaged and fresh-made products. The store was revolutionary in that it eliminated the tradition of checkout counters and cashiers. Instead, a sophisticated mix of cameras and sensor pads recorded every product package a shopper would pick up—or put back—and messaged the shopper with a digital receipt within minutes of walking out the door. They simply left the facility with their desired items, and the store tracked and charged them for what they left with.

In late August 2022, only four and a half years later, Amazon opened an Amazon Fresh store in Chevy Chase, Maryland, featuring the Just Walk Out technology. The store is approximately 46,000 square feet in size.

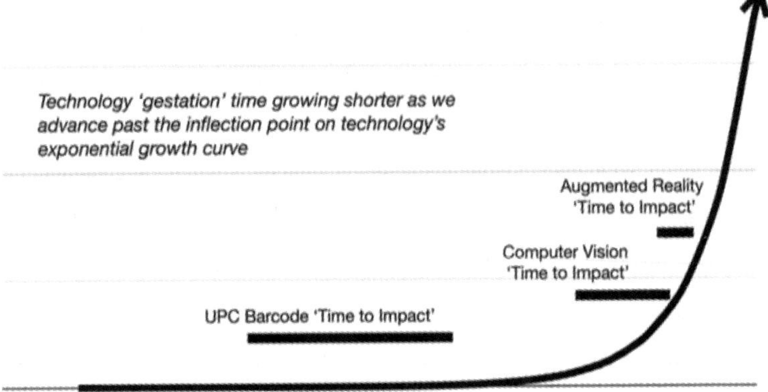

In *Computer Vision Platforms: Seeing the Future of Retail*, I projected that a fully built-out computer vision platform could generate a financial benefit of $75 million for every $1 billion in sales. That's a 7.5 percent financial gain between cost savings and revenue increases; in supermarket retail, with its ~2 percent net profit margin, that's a remarkable number. Now consider the impact Amazon is achieving as it scales deployment of its Just Walk Out technology in an ever-growing number and size of stores. That's why understanding the concept of technological gestation becomes so important.

And it's not just Amazon pursuing this technology. Trigo is an Israeli company growing fast in the space, working with retailers like

Tesco, Aldi Nord, Rewe, Shufersal, Auchan, Wakefern, and others. AiFi, an AI company providing autonomous retail solutions, has perhaps the largest global presence, working with retailers in North America, Europe, Asia, UAE, and Australia. Grabango is another company bringing computer vision to retail, working with Giant Eagle, Circle K, and others.

So, for this technology we observe a gestational period of about eight years, from the first generation of video systems in 2010 to the implementation of the first computer vision platform, Amazon's original Go store, in 2018. And then, in the next four and a half years, a growth of over 2,000 percent in the size of the store powered by computer vision. Time to impact: Let's call it fifteen to twenty years as platforms continue to scale across retail. This is significantly shorter than the decades it took the UPC barcode.

> *"The future will be far more surprising than most people realize, because few observers have truly internalized the implications of the fact that the rate of change itself is accelerating."*
>
> —*Ray Kurzweil,* The Singularity Is Near

As we move up and out the exponential growth curve of technology, the gestation period becomes shorter and the time to scale, or impact, becomes shorter. That means that failing to act early on makes it increasingly difficult for slower movers to catch up to the leaders. Think of it this way: the "slow growth" period becomes shorter as technology speeds up, making the margin for error ever smaller.

That's why it is so critical for retailers to discover, assess, and invest in new disruptive technology early on, not follow, as most retailers have been prone to do for the past century.

The time to react to competitors deploying new transformative and disruptive tech grows shorter as we move into the future. Keep that in mind as we take a look at a transformative technology that's rapidly approaching: augmented reality.

Augmented Reality

I can remember being in Mountain View, California, in the fall of 2016 and seeing all these people walking around the park in the center of town holding up their smartphones in front of them. Commenting on it, my daughter-in-law told me about Pokémon Go, the augmented reality game that had just been released. Wanting to learn more, I downloaded the game and was off and running, searching out virtual Pokémon wherever I traveled. The app coordinates with GPS to allow players to track, locate, capture, train, and battle virtual Pokémon creatures with other users that appear to be in the users' actual locations through phones' camera functions.

Pokémon Go was, for many people, their first exposure to augmented reality, overlaying the real world with digital information. Today, AR is increasingly being used in industry, from manufacturing plants to healthcare, as workers and doctors use specialized headsets able to provide a digital overlay to their physical experiences. For example, Augmedics provides surgeons an augmented reality system that enables them to see patient's anatomy as if they had X-ray vision. A special headset provides a retinal display that guides surgeons during intricate spinal surgery. Ox provides AR-powered headsets enabling warehouse workers to efficiently locate and verify the next product to be picked.

As talk of the Metaverse grows, augmented reality and virtual reality are gaining momentum. And while virtual reality is becoming increasingly realistic, driven by computer processing power and improved graphics, I believe augmented reality is going to be more impactful over the next decade or so. The idea of using a digital overlay on the real world around us to provide relevant information is simply too compelling. Imagine the ability to bring the brick-and-mortar store to life, creating a truly immersive shopping experience, for each individual shopper entering the facility. The opportunities for enriching shopper experiences and building stronger brand loyalties are limitless!

Statista projects there will be over 1.7 billion AR users by the end of 2024, and while the vast majority of them will be experiencing AR through their smartphone, that's about to change.

Apple's Vision Pro, announced recently and scheduled for release in early 2024, is a harbinger of the coming change in how we interact with the world around us. The Apple Vision Pro headset seamlessly blends digital content with the physical space around the user. Given Apple's track record for disrupting new spaces, the Vision Pro is set to usher in the age of spatial computing—a technology that allows computers to blend with the 3D world in seamless, natural ways.* And there's more: while the Vision Pro will be first out of the gates, Apple recently filed patent applications for the next generation of

---

* Ben Dickson, "What Is Spatial Computing? A Basic Explainer," PC Mag, Jun 15, 2023, https://www.pcmag.com/how-to/what-is-spatial-computing-a-basic-explainer#:~:text=Spatial%20computing%20is%20a%20technology,headset%2C%20the%20Apple%20Vision%20Pro.

smart glasses. Today, the iPhone is the hub for Apple Watch and AirPods; smart glasses will be the next addition to the powerful ecosystem Apple has created.

Unlike the first generation of smart glasses (like Google Glass), next-generation glasses will use the processing power of the individual's smartphone and then wirelessly convey the relevant information to display on the lens. This approach gains several things:

- By removing the processor from the glasses frame, it allows next generation glasses to more closely resemble the glasses countless people already wear.
- Smartphones have enormous processing power that can be efficiently utilized to power immersive shopping experiences using this new technology.
- The next generation of smart glasses can leverage a massive installed base of technology—shoppers smartphones and watches. Apple reported in February 2023 that it has more than 2 billion active devices in the market.
- Smart glasses leverage advances in networking and communication speeds to relay relevant information from the retailer to the smart device and on to the glasses.

In a sign of increased market activity, Google, Samsung, and Qualcomm are partnering to bring an augmented reality platform to market and compete with Apple.

Shopping is the intersection of physical, digital, and money, so retail will be disproportionately disrupted by these new AR platforms. This makes it imperative for retail leaders to understand, embrace, and implement AR technology to expand consumer experiences, sooner rather than later.

In addition to the requisite technologies, it will be the value proposition to the consumer that drives the explosive growth in AR, and retailers are ideally positioned to fuel this value proposition and provide a truly immersive, powerful, and relevant shopping experience for each individual shopper.

Imagine greeting a customer by name digitally as they enter a grocery store. In one lens is their shopping list, along with wayfinding directing the shopper to the nearest item; think Waze for the store. As the shopper moves into the store, products offering personalized promotions light up. Products that align with and support the shopper's health profile are highlighted, making them easier to find. Walking through the produce department, a window pops up with a short video about the local farm supplying sweetcorn to the store. Or in the seafood department, a video appears providing information on the sourcing and sustainability of the salmon.

And then there's AR for store associates. Imagine store associates wearing smart glasses as they go about their work. The technology automatically recognizes out-of-stock items as the associate walks down the aisle, or notices that a display of bananas is becoming overripe. Or the online order-picker being directed to the next product, and then the tech automatically validating it as the correct item. AR not only improves the consumer experience, but it makes store operations more effective and efficient.

AR-powered smart glasses also have the potential to disrupt the nascent frictionless checkout capabilities. Imagine each time a shopper picks up a product while they're shopping, it just automatically scans and is added to the transaction, without requiring them to go through a traditional checkout process. AR-powered shopping provides myriad advantages over ceiling mounted or shelf mounted cameras; the consumer (shopper) provides the hardware, keeps it updated, and uses a camera within a couple feet of the product being purchased. As you get a sense, there is almost no limit to the applications for augmented reality smart glasses in retail.

Like in our previous examples for the barcode and computer vision platforms, it is the convergence of exponentially growing technologies that is powering up augmented reality. Like every application, growing computer processing power and trusted data are the fundamental enablers. But it is processing power combined with higher network communication bandwidth, digital lenses, and AI-powered image recognition that is setting the stage for explosive growth.

Given the convergence of all the required technologies, a massive installed base of technology (smartphones) to build on, and a powerful value proposition and business case, the gestational period for AR in retail is going to be much shorter than the barcode or computer vision platforms.

But here's the catch and why an understanding of technology's gestation becomes so important: augmented reality in retail needs a digital foundation, a digital twin of the physical store and its products.

Building that digital twin takes time. Forward-looking retailers are already beginning to build that digital foundation in preparation for what is to come. And the retailers who are ready to capitalize on the availability of consumer smart glasses as they are released in the next twenty-four months will provide a powerful value proposition to shoppers, driving loyalty and new shopper acquisition. Slower-moving retailers will be at a loss, unable to compete until they build their digital twins across all their stores and products, create their platform, and build a value proposition.

Ross Finman, leveraging significant experience in the space gained through his time with Niantic (producer of Pokémon Go), is working hard to make this AR vision a reality. His company, Augmodo, understands the need for retailers to begin building the digital foundation today to bring AR shopping to life tomorrow. Augmodo today helps retailers create an augmented workforce, to begin building a digital twin of the store.

Creating an accurate digital replica of the store, including all the products, provides the needed foundation for augmented and virtual reality experiences. Augmodo builds this digital twin by outfitting store associates with "smart" ball caps that have tiny, nearly unnoticeable, cameras built into the brim. As the associate goes about their job stocking shelves, the cameras are recording the layout of the store, including products and their location.

Alternatively, Augmodo provides a system for picking online orders. The order picker wears smart glasses having a built-in camera. The glasses guide the worker to the next product to be selected and automatically verify the correct product was picked. All the while, the camera is recording store layout and building up digital versions of all the products across the store.

Augmodo is digitizing the store, and all the products within it, as the company provides growing services to retailers and CPG brand manufacturers. As Ross says, "you need to understand reality in order to augment it."

So, we have now entered a world past the inflection point on the exponential growth of technology. The innovation flywheel—exponentially growing processing power, data, and AI—is spinning ever faster, launching a constantly growing number of innovation butterflies into the market. And those butterflies are achieving scale at an ever-growing pace. And any one of those butterflies can cause major disruption in your business.

Amidst this chaos of tech-fueled innovation, retail executives must continue to operationally execute, satisfy shoppers, grow sales and margins, and control costs.

What else must retail leaders do? **Shift focus.**

Chapter Summary

- Retailers have less and less time to discover, test, and deploy new technologies. Begin to normalize quick iteration and become comfortable with the idea of failing fast when piloting new technologies so the expectations to move through ideas quickly is not as big of a shock to the system.
- Remember, in the post-inflection world, speed doesn't kill—being slow does.
- A digital duplicate of your retail environment is needed to implement a growing array of new technologies. Consider partnering with tech firms to begin mapping your store space and experience, and at the very least start to understand how your space could be augmented by these burgeoning technologies.

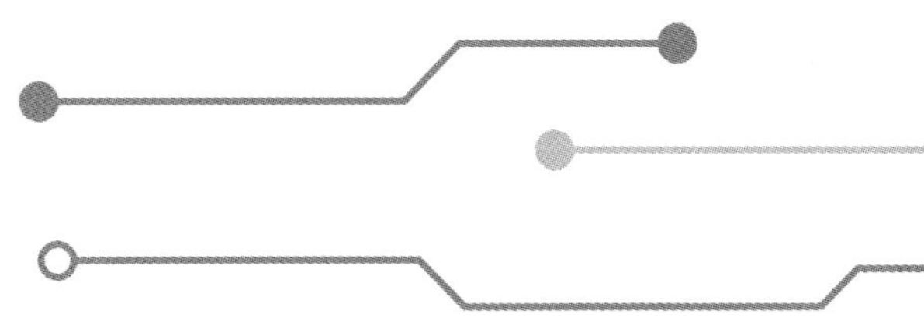

Part 2

RETAIL IN AN EXPONENTIAL WORLD

"The future depends on what you do today."

—Mahatma Gandhi

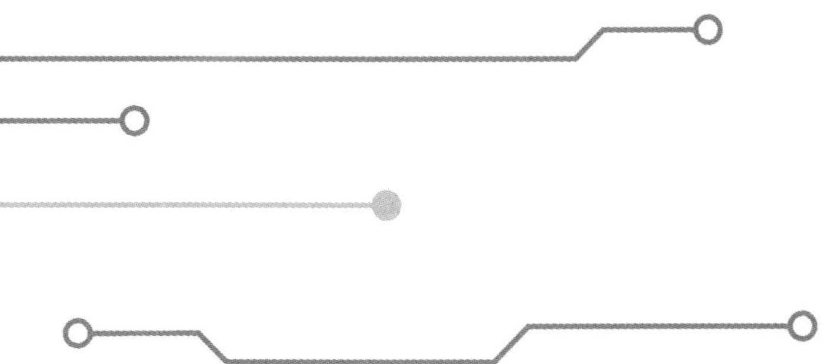

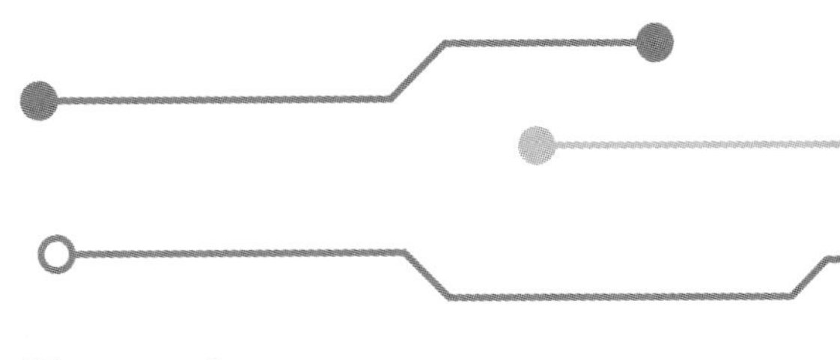

Chapter 4

SHIFTING FOCUS

X-rays don't lie. A couple years ago, I was in an orthopedic doctor's office, driven there by excruciating knee pain—pain so bad I could barely walk. The problem had started several weeks before when we were moving to be closer to family. A few days of moving heavy boxes had quickly taken a toll on my knee, but given we were on a timetable for the move, I soldiered on. Tylenol and Aleve became my best friends.

So, there I was in the doctor's office, waiting to hear the diagnosis and looking forward to getting some relief. The doctor came in the room, pulled up the X-rays, and told me my knee looked fine. But how was my hip? My hip? My hip was good, not bothering me at all. It turns out that I had developed osteoarthritis in my hip, and the pain referred to my knee, which it sometimes does.

The X-rays didn't lie; they clearly showed the deterioration of cartilage in my hip joint. It was my mind that had been "lying" to me, making me think the agonizing pain in my knee was the problem, when, in fact, the real issue was something entirely different.

I speak with executives from across the retail industry every day, and the conversations inevitably focus on the chaos of new innovation. What new capability should they be concerned about? What new digital-native competitors should they be watching? What are shoppers expecting? But retail executives are focused on the wrong things. Just like I was with my knee.

It is only by looking past the overwhelming chaos of new innovation that retail leaders can begin to understand that all these new technologies are simply manifestations of a far more serious, fundamental change in the world. Technologies come and go at an ever-increasing pace. Only by understanding that we now inhabit a world that has never before existed—and that operates by fundamentally different rules—can executives begin to prepare themselves and their organizations to meet the new challenges.

Imagine being able to simply think about browsing the web, sending an email, or even shopping online, and it happens. That's just what Elon Musk's company Neuralink is focused on: developing a brain implant that connects the human brain directly to a computer. Synchron, a competitor, has already beat Musk to that milestone by implanting a device into a patient with ALS, a neurodegenerative disease. Synchron's Switch and its accompanying array of sensors inserted into the brain via a blood vessel give the patient the ability to send messages through WhatsApp and make online purchases—all by thinking.*

---

* Alena Botros, "Elon Musk's Neuralink Brain Computer Startup Is Beat Again. This Time a Competitor Implanted Its Device into Its First U.S. Patient," Fortune, July 18, 2022, https://fortune.com/2022/07/18/elon-musk-neuralink-beat-by-synchron-brain-computer-startup-us-human-trial/.

This isn't some far-off technology or scene from a science-fiction movie. This technology is here. Now. Think about the impact on shopping when all you have to do is *think* about purchasing something.

And the challenges for traditional retailers are becoming more significant by the day. Technology now drives business strategy, something that is natural for digital-native companies but very foreign to brick-and-mortar operators. Retail competition is transforming, driven by big data, artificial intelligence, and digitalization—far from the historical norm of product-based mass price promotion. How value is created is opening up into new dimensions, far beyond selling that package on the shelf. And people problems are just getting started. Are you ready to train and re-train your workers at an ever-faster pace as new capabilities enter the marketplace? Or perhaps future retail doesn't have many human workers, using computer vision systems, physical automation, and AI-powered business process automation to perform tasks and make decisions across the enterprise.

Now Entering the *Twilight Zone*

One of the most famous episodes of *The Twilight Zone* is "Mirror Image," which aired in the show's debut season in 1960. In that episode, a young woman named Millicent Barnes is waiting at a bus station for a trip to Cortland. She notices that her suitcase is missing and, while searching for it, goes to the restroom to freshen up. When she returns, she discovers that her ticket has been changed and her seat has been taken by someone who looks exactly like her.

Millicent becomes convinced that her doppelgänger, who appears to be attempting to take over her life, is stalking her. She confides in a kind stranger, who tries to reassure her that she is simply imagining things. But as strange occurrences continue to happen, Millicent begins to suspect that she has entered a parallel universe where everything is almost the same but not quite.

Go into your stores and look around. Brick-and-mortar stores look about the same as they did yesterday. Products on the shelf, shoppers flowing in the door and up and down aisles. But look a little harder, and you'll see subtle differences. Shoppers look at the microcomputers held in their hands or on their wrists as they shop, checking lists, nutrition values, ingredients, and sourcing. The shopping carts have digital screens, silently tracking where they (and the shopper) go in the store. The shelf tags are digital, changing prices in real time driven by demand, inventory, and even the weather. In some stores, the checkout is gone, replaced by a spider's web of cameras and sensors across the ceiling.

And that's just what you're able to see. Behind the scenes, artificial intelligence agents are monitoring inventory, ordering products, and tracking deliveries to massive, dark, automated distribution centers. Robots are picking cases, automatically building pallets for each aisle of the store they're being sent to, without a worker in sight. Self-driving delivery trucks bring the goods to the store, coordinated by other AI agents. Automated fulfillment centers pick orders placed online; self-driving delivery bots trundle down the sidewalks, making home deliveries. You get the idea. Almost the same, but not quite.

And maybe you'll even see Millicent shopping in the aisles.

Inexorable, Exponential Growth . . .

Understanding this new otherworld begins with understanding exponential growth. Like gravity, the exponential growth of technology is pervasive, a near-universal force impacting every part of the massive retail industry, and that impact is growing greater every day.

The word "exponential" is bandied about quite frequently today, and often without rigor, when referring to advancing technology. To many people, exponential simply means fast. But true exponential growth accelerates, and that's what makes it so difficult to wrap our heads around. And yet it is so vital for retail leaders to grasp the true meaning of exponential so as to make sense of what lies ahead.

Here's a quick exercise that shows the power of exponential growth. Take a moment and get out a sheet of paper. Lay the sheet of paper on a table and fold it in half. Then fold it in half again. A typical piece of paper is around 0.004-inch thick. If you were to continue folding that piece of paper fifty times, how high do you think it would be? Ten inches? A meter? More?

The answer: approximately 95 million miles—the distance from the Earth to the Sun.

That's the power of exponential growth.

As we read earlier, computer processing power has been growing exponentially for decades; the difference is that we have passed the inflection point and technology-driven innovation has burst forth from the labs. It has become noticeable. And it is happening everywhere.

A Double-Barreled Threat

Exponential technology growth brings with it a double-barreled threat. Failure to recognize exponentially growing technology early on is the first threat. We have all experienced firsthand the impact that exponential growth can have. As we saw in the first days of the pandemic, the spread of the COVID-19 virus happened so slowly and in such small increments that it didn't merit much attention. And that's just what made it so dangerous. Virologists clearly understand the danger of exponential growth, which is why such effort is put into stopping viral spread early on. Past a certain point, it's nearly impossible to stop. But while virologists get it, ordinary laypeople do not.

It is human nature that makes the notion of exponential growth so difficult to truly comprehend . . . and that's what makes the time we are in so treacherous for retail leaders. When we project out into the future, we intuitively assume the current rate of change will be maintained. But it is the inexorable, exponential, and accelerating growth of computer processing power that underpins the massive influx of innovation happening today. And the pace of change will be faster tomorrow, and then even faster the day after tomorrow. It is constantly accelerating. And that is a foreign concept to us humans.

Kodak didn't recognize the menace of digital photography until it was too late, destroying a century-old corporate icon. Blockbuster didn't recognize the dangers of internet-based streaming until Netflix destroyed the DVD rental business. The music industry didn't appreciate the peril of digitized music, personified by the iPod. And those examples are now years old, when technology was moving much more slowly. Today, it's a fair bet to say that traditional retailers are underestimating the power of artificial intelligence.

The second barrel of the threat is a failure to appreciate how fast technologies can scale past the inflection point, causing widespread disruption at eye-watering speed.

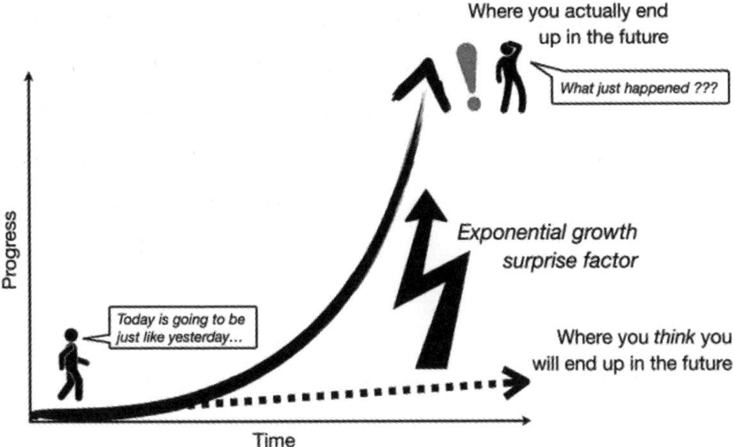

Exponential growth bias is a phenomenon whereby humans underestimate exponential growth. In the world of retail, this appears as executives thinking it will take far longer for a given technology or capability to impact their business than it actually will. Human beings are hard-wired to think linearly, and that's the basis for our intuitive decisions about what is going to happen when. From experience, we intuitively know when to catch a ball thrown to us, how long it will take to build a new store, or to pay off a loan. But what happens when our intuition misleads us? When our linear mindset becomes increasingly hazardous as we proceed into the future?

Failure to Act Can Be Fatal

It's easy to dismiss the threat of exponential growth early on. That's what makes failing to understand what's happening with technology so dangerous today. And that's just the mistake traditional retailers are making. Failing to act during the slow growth period makes it increasingly difficult to catch up to those who are riding the exponential growth curve of disruptive technology into the future. *Our minds think in linear, not exponential, terms. Thirty linear steps will get you about 30 meters. Thirty exponential steps will take you twenty-six times around the Earth. If I gave you $1 a day for thirty days, you would have $30. If I gave you 1 cent but doubled it every day, in thirty days you would have $10 million.*[*]

It is that understanding of the nature of technology growth that makes digital-native companies dangerous to traditional organizations. In the food industry, for example, nearly every brick-and-mortar retailer I speak with dismisses Amazon as a significant near-term or even mid-term threat. They have a limited number of physical stores. They haven't figured out logistics and inventory. They have out-of-stock issues. They aren't turning over fresh food fast enough. Don't be fooled: Amazon will figure out food retailing. What traditional retailers should be more scared of is Amazon's technological prowess to create the future of retail. Amazon doesn't wait or avoid exploring new innovation. They quite literally ride the wave and sometimes even make it themselves.

According to the *Wall Street Journal*, "[Amazon] has a long

---

[*] Peter H. Diamandis, "What Does Exponential Growth Feel Like?" Diamandis.com, December 5, 2017, https://www.diamandis.com/blog/what-does-exponential-growth-feel-like.

history of solving the thorny problem of getting consumers and their stuff into close proximity. It also has a new boss with a powerful motivation to rejuvenate the company's sales growth, plus an estimated 167 million members of its Prime shipping service in the U.S. alone, according to Consumer Intelligence Research Partners, who are already inclined to concentrate their spending on the company's offerings."*

A case in point: Amazon recently installed the Amazon One palm-reading payment system in our local Whole Foods store. They've built a process and supporting technology to map unique features of individuals' palms and use it to authorize purchases by hovering your hand over the designated scanner. The seemingly small act of having to take out my wallet, find a credit card, insert it, or wave it over the payment terminal has now become a massive headache and time-suck now that I can just wave my hand over the reader.

Contrast that ease with the miserable self-checkout experience encountered by shoppers at the King Sooper (Kroger) store down the street, where the screen inevitably freezes multiple times during checkout with a message of "Help is on the way," though it may take several agonizing minutes before help arrives to clear the issue.

Checking out at King Sooper gives me a headache. Checkout at Amazon Whole Foods has removed a headache and made my shopping that much easier. And not just at the Whole Foods near me, Amazon recently announced it is rolling out the palm payment system to all its stores across the United States, in addition to select

---

* Dan Gallagher and Jinjoo Lee, "Why Amazon Isn't Checking Out of Groceries," *Wall Street Journal*, May 12, 2023, https://www.wsj.com/articles/amazon-groceries-fresh-go-13f27a93?mod=hp_lead_pos6.

Panera restaurants, airport retailers, and sporting arenas.

Retail's Wheels Are Coming Off

"Our biggest companies and government agencies were designed in another century for purposes of safety and stability," explains Peter Diamandis in *The Future Is Faster than You Think*. "Built to last, as the saying goes. They were not built to withstand rapid, radical change. This is why, according to Yale's Richard Foster, 40 percent of today's Fortune 500 companies will be gone in ten years, replaced, for the most part, by upstarts we've not yet heard of."[*] If nearly half of the Fortune 500 companies are going to be disrupted out of business, what does the future bode for traditional retailers?

The retail industry is structurally ill-prepared for what lies ahead. Company culture, associate skillsets, systems and processes, resource allocation, leadership styles, organization structure, supply chains, and business models were built for the linear-paced world of yesterday and are increasingly misaligned with the exponential-paced world of today, let alone tomorrow.

And it is retail executives themselves who are perhaps the most significant challenge. For decades, executives who started their careers stocking shelves or bagging groceries have been lauded for earning the respect of their fellow associates by coming up through the ranks. But the mindset caused by this career path creates its own set of problems in the new, increasingly digitized world of retail.

---

[*] Peter Diamandis, *The Future Is Faster than You Think* (New York: Simon & Schuster, 2020).

For most executives, this kind of career path reinforces the "we've always done it this way" approach. Unknowingly, many retailers pick up what become limiting certitudes as they move through their career that create significant hurdles when it comes to innovating the future.

Doug McMillon, CEO of Walmart, is a rarity in retail, a lifelong retailer who has been able to understand the implications of technology's growth and transform the world's largest retailer into a hybrid traditional–digital enterprise. Though Walmart's journey has had some twists and turns, McMillon is to be credited for his early vision of how technology would transform traditional retail, his commitment to investing in new innovation, and to setting aside outdated practices.

We are rapidly approaching a point where traditional retail, in its present form, cannot keep up with ever-faster disruptive innovation, new digital-native competitors, and rapidly shifting shopper expectations. Like the proverbial car accelerating past its mechanical limits, retail's wheels are starting to come off, and a growing number of companies are headed for a crash.

And yet, too many retail executives aren't facing the future. Any discussion around the implications of doing business in this new otherworld past the inflection point is notable by its absence. As I talk with industry executives, even executives at the big consulting companies that are supposed to be helping retail move into the future, there is little to no discussion around how organizational structures need to change. Or how long-standing decision-making processes need to be completely restructured to be effective in an ever-faster world. Or how resources need to be reallocated to support early investment in game-changing capabilities. Or how long-established business models are being disrupted.

As a growing number of companies in the retail industry are beginning to realize, the cost of not confronting the world as it is today can be significant. Lost sales, declining margins, defecting shoppers, a slow decline into irrelevance, or even abrupt business failure.

Too many retailers are dealing with the symptoms of innovation rather than addressing the underlying forces. And this approach is dangerous, leading executives to make short-term tactical decisions without understanding the larger battle being waged. Consumed with the nuts-and-bolts tactical execution required to simply stay alive in such an intense, high-volume, low-margin business, retail leaders are challenged to step back and understand the vast forces at work.

And yet they must understand the need to change paths if they are to succeed in the new world of retail.

Which path will you choose?

Chapter Summary

- Retail leaders are faced with a double-barreled threat in the wake of exponential technology growth: failure to recognize exponentially growing technology early on and failure to appreciate how fast technologies will scale past the inflection point.
- Early on, exponential growth bias causes humans to underestimate the power and impact of exponential growth yet this understanding is vital to guiding your organizations in a world of accelerating technology.

- Don't miss out on your opportunities to explore, understand, and implement new technologies. Failing to act during the slow growth period makes it increasingly difficult to catch up to those who are riding the exponential growth curve of disruptive technology into the future.

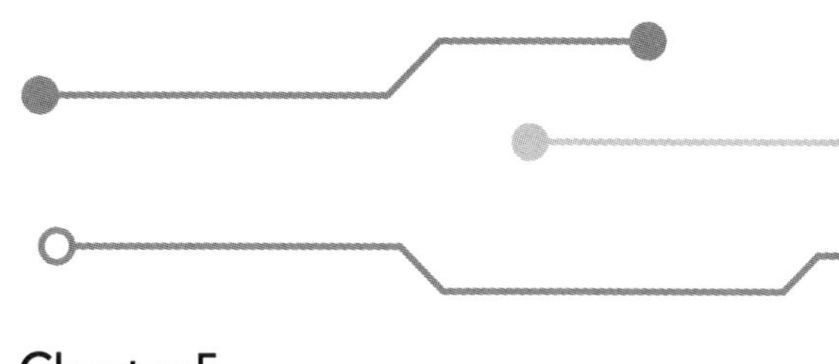

Chapter 5

CHANGING PATHS

Caesar was a prominent Roman military general and statesman, who played a crucial role in the transformation of the Roman Republic into the Roman Empire.

In 49 BCE, Caesar found himself in a power struggle with the Roman Senate and his political rival, Pompey the Great. The Roman Senate ordered Caesar to disband his army and return to Rome, as they feared his growing influence and popularity. However, Caesar was hesitant to comply with the Senate's orders, knowing that he would lose his protection and potentially face prosecution for his actions as a military leader.

Caesar found himself at a point of no return when he reached the Rubicon River, which marked the boundary between Gaul (modern-day France) and Italy. According to Roman law, a general was forbidden from bringing his army across the Rubicon into Italy without explicit permission from the Senate. Crossing the Rubicon with his troops would be seen as an act of rebellion and would be considered treason.

Contemplating his next move, Caesar famously declared, "Alea iacta est" (The die is cast) and made the momentous decision to cross the Rubicon with his army. By doing so, he knowingly violated Roman law and set in motion a civil war against the forces loyal to the Senate.

Caesar's decision to cross the Rubicon had profound consequences. It marked the beginning of the Roman Civil War, with Caesar leading his army against Pompey and his supporters. The conflict lasted for several years and resulted in Caesar's ultimate victory, solidifying his power and authority.

The crossing of the Rubicon was a significant turning point in Roman history. It shattered the traditional norms and legal boundaries that had governed the Roman Republic and set the stage for Caesar's consolidation of power. Ultimately, it led to the downfall of the Roman Republic and the establishment of Caesar as the first dictator of Rome, marking the beginning of the Roman Empire.

The phrase "crossing the Rubicon" has since become a metaphor for making an irrevocable decision, symbolizing the point of no return or the moment of taking a decisive and irreversible action.

You have a decision to make.

Retail's river of innovation is growing wider, and the water is rushing ever faster, making it more difficult to cross every day that goes by. The battle to create the future has already been joined; Caesar has crossed the Rubicon. Amazon, Walmart, and Kroger are all investing massive resources to create their retail futures.

Walmart is investing in Symbotic's distribution center automation, with the company claiming that unit cost averages will fall by some 20 percent.*

The largest retailers are creating their own retail media networks (RMNs), creating significant new revenue sources. In 2022, Walmart generated ad revenue of $2 billion, Target generated $1 billion;† Kroger's Precision Marketing retail media arm generated $1.2 billion in operating profit in 2022.

Kroger is leveraging shopper-identified purchase data from 60 million households and 2.8 billion annual visits across 2,800 stores in thirty-five states to shift marketing to precision-targeted promotions while killing the traditional print weekly ad in a growing number of markets. Kroger is seeking to acquire Albertsons, not to lower the cost of a product by a penny or two, but to have the resources to compete with Walmart and Amazon to create the future.

The traditional retail industry has reached its Rubicon, an inflection point demanding a decision. It is time for you to decide if you are going to change paths, making a conscious choice to leap to the exponential path and compete in the post-inflection point world or to remain on the linear path that you've lived.

Changing paths to participate in an exponential future requires deliberate action. And failing to make a decision to change paths has consequences.

---

* Daphne Howland, "Walmart Banks on Stores, Robots as It Stokes e-Commerce," Grocery Dive, April 5, 2023, https://www.grocerydive.com/news/walmart-automated-stores-robots-e-commerce/646918/.

† James Hercher, "Why 2022 Was A Year Of Reckoning For Retail Media," Ad Exchanger, December 29, 2022, https://www.adexchanger.com/commerce/why-2022-was-a-year-of-reckoning-for-retail-media/.

Retail Purgatory

Not making a conscious decision, just continuing to go along, adopting whatever new technologies you're forced to or that you see your competitors doing, is, in my mind, the worst thing you can do. Refusing to consciously pursue one path or the other will commit you to retail purgatory.

Technology and tech-powered innovation are happening too quickly, too pervasively, and too disruptively to not be deliberate about them. Yet we see far too many retailers not developing a strategy—or, more importantly, a mission—to guide their actions into this new frontier. And as we read about in Chapter 4: Shifting Focus, it's not about this or that new technology, it is understanding that we now live in a world unlike any before, a world of exponential technological growth, and that changes everything. We need to look at these changes as a sweeping, universal shift in how we experience the world around us.

In short, creating the future takes work. And if you're not ready to commit yourself and your organization to that work, then save the effort. Just continue along as you've been. But know that others are going to create the future—maybe some other traditional competitor or maybe some new digital-native company—and at some point, you're going to fall behind and be increasingly out of sync with what your shoppers want. You will likely become irrelevant. Maybe that decline won't happen for years. Or maybe it will happen soon. Given the pace of new innovation, I'd bet on it sooner rather than later.

Once a dominant retail powerhouse, Sears failed to adapt to the rise of e-commerce and shifting consumer preferences. The company struggled to invest in online retail and digital innovations, allowing competitors like Amazon and Walmart to gain a significant advantage. As a result, Sears faced declining sales, store closures, and filed for bankruptcy in 2018.

Crossing the Rubicon: Change Is Hard

Change is always easier when what you've been doing no longer works. So, in that regard, this industry inflection point demanding a choice does not come at a good time. Retail, especially consumer packaged goods (CPG) retail, has been riding pretty high, coming off strong pandemic sales and, more recently, inflation-driven sales and margins. There's no pain.

And that makes it really hard for leaders to make the decision to change paths, to move away from comfort to something that's going to be uncomfortable. People naturally resist change, especially when it disrupts established routines and ways of doing things and when "nothing feels broken." Employees may be comfortable with the way things are and may fear the uncertainty and potential risks associated with a new strategy. Overcoming resistance and fostering a culture of openness to change can be significant hurdles to overcome. We'll discuss supporting organizational change management and mindset shifts in Chapter 6: Retail Transformation.

Organizations also develop a certain inertia or resistance to change over time. Corporate structures, processes, and systems evolve to support and enable the business to operate in a certain way. This inertia can make it difficult to implement new strategies, as it requires disrupting established practices and overcoming bureaucratic barriers.

Creating the future involves risk and uncertainty. There is no guarantee that the new path will be successful, and the fear of failure can make decision-makers hesitant to embrace change. This fear can be amplified in risk-averse cultures or industries where the consequences of failure are perceived to be high. Unlike tech companies, traditional retail companies shun risk unless their back is to the wall.

Kroger is an exception to this. Zach Bello was part of Kroger's innovation team when the company initiated a pilot in late 2017; they tried using Nuro's self-driving vehicles to deliver online orders. As Zach calls out, "Kroger leadership actually blessed risk-taking, knowing full well that the Nuro pilot may not be successful. But leadership realized that even if self-driving vehicles were not yet ready for prime time, they would be at some point in the future, and Kroger would be positioned to leverage the learning from the pilot to gain an early-mover advantage."

Businesses may be reluctant to change strategies due to the sunk cost fallacy—the tendency to continue investing in a failing course of action simply because significant resources have already been expended. Leaders may be unwilling to abandon previous investments, even if they are no longer aligned with the organization's new objectives. But sometimes investing in new, risky paths provides an immeasurable payoff through the knowledge gained to be applied to later technologies. Just as we've discussed before, growth in tech is usually iterative.

Changing paths to a new strategy requires a clear vision and direction for the organization. Without a well-defined plan and a compelling rationale for the change, stakeholders may struggle to understand the need for a shift and may resist the proposed strategy.

Organizations are complex social systems, and internal politics can hinder strategic change. Individuals or groups with vested interests in the status quo may resist or actively oppose changes that threaten their power, influence, or established ways of working.

> *Walmart's CEO Doug McMillon shares the ups and downs of Walmart's journey: "We wish we had been more aggressive early on, no doubt. In some ways we experienced what Clay Christensen calls the 'innovator's dilemma,'" said McMillon. "We hired talent, invested, and just kind of meandered along rather than hammering down, being aggressive, and making it a must-win aspect of our business. That's partly because we had a bird in hand. We knew that if we continued to open Walmart Supercenters, they would do well. Traffic in the United States is still going up. But digital conversion for us has to be about more than just serving the customer on the front end. It's about more than e-commerce. We need to introduce digitization across all our functions and jobs so that we can be faster and more efficient."*
>
> *McMillon is describing the biggest risk to any organization—not fully embracing change for fear of disrupting existing business.**

Your Inflection Point

As you and your organization prepare to cross the Rubicon—or not—think of the decision as your own inflection point. The choice is

* Mackenzie Johnson, "CEO Doug McMillon Discusses Innovation and the Future of Walmart," ActionIQ, accessed September 2023, https://www.actioniq.com/blog/ceo-innovation-future-walmart/.

simple: you are either going to consciously and deliberately change to succeed in the exponential world or you are not.

Which path will you and your organization take?

Chapter Summary

- The traditional retail industry has reached its Rubicon, and it's time for retail leaders to make a decision regarding their organizations' futures: make a conscious choice to leap to the exponential path and compete in the post-inflection point world or remain on the linear path that you've lived.
- Committing to a future in the exponential world carries with it inherent risk and uncertainty. Overcoming a history of slow adoption of new capabilities is increasingly imperative for survival, let alone success. Becoming nimble in regard to testing and deploying new technologies along with becoming comfortable with fast failing are necessary as we move into the future.
- Deliberate consideration and strategy are imperative partners to agility and quick change adoption. Merely flipping through new technology capabilities without a plan or integrated approach will not reap sustained growth or stability.

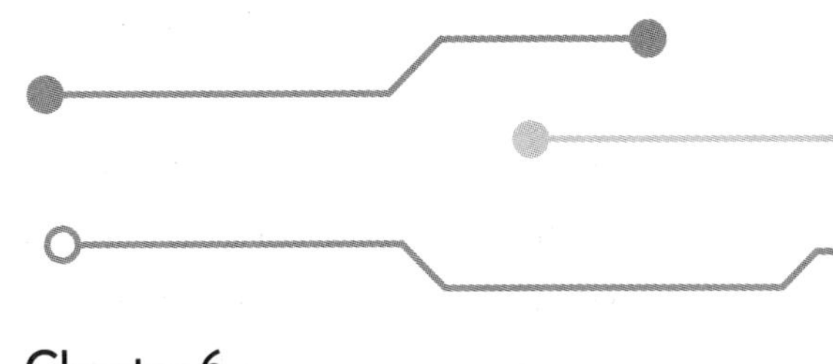

Chapter 6

RETAIL TRANSFORMATION

So, I assume you've made the decision to create your retail future. How do you do that? Where do you start?

Changing paths, deliberately shifting from the past linear path that leads to retail purgatory to the new exponential path requires work. It requires doing things differently. It requires creating a new culture, leaving restrictive beliefs behind while creating new beliefs around what is possible, new organizational structures, and new ways of thinking and doing. It requires being uncomfortable.

The most important part of transforming yourself and your organization for this new world is not technology. As we read earlier, technologies come and go at a dizzying pace. Sure, we have to remain agile and functionally sound enough to understand these new tools and implement them, but that's not the deeper challenge facing us. Retailers need to look beyond specific technologies to the bigger picture—the new world in which we now live—and prepare their organizations to thrive in a new environment. Leaders need to influence their organizations in a way that empowers them to not only manage the accelerating changes we're encountering, but also to change mindsets to actually embrace the constant changes. Success in this post-inflection point world requires changing your beliefs around what can be. It's being open to new possibilities. It's using inspiration to trigger your own inflection point to jump from one path to the other. And then it's realizing the potential you have to create the future in a world where most anything is becoming possible.

And I'll warn you ahead of time, the discussions around changing beliefs and new possibilities are challenging, as I can readily attest to. I'm a baby boomer who spent many years as a retailer and have a pretty strongly established view of how things work and how I view the world, so in many ways a prototypical retail executive. As many retail executives can attest, as you grow through your career, you're able to leverage the different experiences you've had over the years, bringing that learning to bear on new situations. The challenge is that technology is creating a world that has never before existed, and knowledge gained through our past experiences is sometimes not as relevant.

My son Sterling is an author and top-rated motivational speaker who helps people break through limitations to achieve their potential. He, along with my daughter Haviland, who works with Sterling, and I, have had countless discussions around what it takes to think differently and let go of past assumptions to change paths. It's not an easy process, but a necessary one, if you're going to successfully navigate the uncharted waters of the future.

Lego's Journey

In 1949, LEGO introduced its first interlocking plastic bricks, a groundbreaking innovation that would lay the foundation for the company's future success. Those bricks, known as "LEGO bricks," allowed children to construct and rebuild endless creations, sparking their imagination and fostering their cognitive and motor skills.

In the decades that followed, LEGO experienced significant growth, becoming a staple in children's playrooms worldwide. The company's commitment to quality, durability, and the limitless creative possibilities of its bricks propelled its success. However, in the late 1990s and early 2000s, LEGO faced a daunting challenge: changing market dynamics, increased competition from electronic toys and digital entertainment, and a lack of innovation threatened the company's relevance and financial stability.

Recognizing the need for change, LEGO embarked on a transformative journey to adapt to the digital age and redefine its place in children's play. This journey was marked by a series of key strategic decisions that shaped LEGO's transformation and enabled it to thrive in the digital world.

One of the pivotal decisions was to embrace digital technologies rather than viewing them as a threat. The company understood that the digital revolution presented an opportunity to enhance the LEGO experience, rather than render it obsolete. In 1997, LEGO launched its first website, marking its foray into the digital realm.

LEGO recognized that children were increasingly engaging with digital media and sought to integrate digital play experiences with physical brick building. In 1998, the company introduced LEGO Mindstorms, a robotics system that allowed children to build and program their own functioning robots. This innovative product combined the tactile joy of LEGO bricks with the exciting possibilities of technology, opening new horizons for creative play.

As the digital landscape continued to evolve, LEGO embraced video games as a means to engage with its audience. In 2005, the company released the first LEGO-themed video game, *LEGO Star Wars: The Video Game.* This marked the beginning of a successful collaboration between LEGO and renowned video game developer TT Games. LEGO video games, featuring popular franchises like Harry Potter, Batman, and Marvel, became a significant part of LEGO's strategy to connect with children in the digital realm while staying true to its core values of creativity and imagination.

In the digital world it's easy to connect networks of users/people across disparate businesses. And as we move into the future, creativity in partnerships becomes increasingly important. Lego's embrace of video games is great example of this. In addition to digital initiatives, LEGO also prioritized partnerships and collaborations to expand its reach and relevance. The company formed alliances with major entertainment brands, such as Disney and Warner Bros., to develop licensed LEGO sets and video games based on popular movies and TV shows.

LEGO's digital transformation was not limited to external initiatives but also extended to internal processes and operations. The company adopted new technologies to streamline its design and manufacturing processes, improving efficiency and speed to market. Digital design tools and advanced production techniques enabled LEGO to introduce more intricate and detailed sets while maintaining the high quality and precision that the brand was known for.

LEGO recognized the power of user-generated content and the rise of social media as platforms for creativity and community building. The company launched LEGO Ideas in 2008, a program that allowed fans to submit their own LEGO designs for consideration. If a design received enough support from the community, LEGO would possibly produce it as an official LEGO set. This initiative not only nurtured a sense of ownership and collaboration among LEGO enthusiasts but also served as a valuable source of inspiration for future LEGO products.

LEGO's digital transformation was accompanied by a strategic focus on expanding its presence in retail. The company invested in its own branded LEGO Stores, creating immersive and interactive shopping experiences that showcase LEGO sets and encourage hands-on play. The stores serve as destinations for LEGO fans, offering exclusive products and events. By maintaining control over its retail outlets, LEGO ensured a consistent brand experience and direct interaction with its customers.

LEGO's transformation efforts have been immensely successful, revitalizing the brand and positioning it as a global leader in play and creativity. The company's revenue and market share have soared, and LEGO sets have become highly sought-after collectibles among both children and adult enthusiasts. LEGO's success can be attributed to its ability to balance innovation with its core values of creativity, imagination, and quality. They saw changes disrupting their market, and instead of refusing to learn and grow, they embraced changes and found ways to incorporate trends and shifts into their brand.

The key lessons that emerge from LEGO's transformation journey are the importance of embracing change, being open to opportunities afforded by technology while staying true to the brand's essence. By recognizing the evolving needs and preferences of its audience, LEGO was able to adapt and thrive in the digital world while maintaining the timeless appeal of its iconic brick-building system. It's important to use new innovative capabilities to reinforce your brand's essence and mission. Never lose sight of your original brand's mission, purpose, and identity, but use innovation to find creative ways to make new meaning and impact.

Transformation like LEGO undertook is not for the faint of heart. Like the caterpillar becoming a butterfly, LEGO retained the DNA that made it special while transforming to succeed in a changed world. LEGO *believed* it could create a new future.

The Power of Beliefs

To help convey the power of beliefs, Plato relates a story about people living in a cave. In the allegory, a group of people are chained inside a dark cave and positioned in such a way that they can only see shadows on the wall in front of them; those shadows are the only reality they know. One of the people is freed and sees the outside world for the first time. He is stunned by the brightness of the sun and the beauty of the moon and stars, and he realizes that the shadows are only a distorted reflection of reality. However, when the freed person returns to the cave to tell the others what he has seen, they reject his claims because their beliefs about reality are so deeply ingrained.

Beliefs influence an individual's motivation and decision-making process. Personal beliefs provide a framework for evaluating options, setting goals, and making choices aligned with one's values and principles. They can provide individuals with a sense of direction, guiding their actions and aspirations and giving them a sense of fulfillment and satisfaction. And beliefs impact an individual's emotional well-being. Positive beliefs can foster optimism, resilience, and a sense of control, while negative beliefs can contribute to stress, anxiety, and negative emotions.

Organizations have collective beliefs that create a company's culture. When people in an organization share common beliefs, it promotes unity, alignment, a common purpose, and collaboration. The beliefs and values upheld by an organization can shape its reputation and how it is perceived by stakeholders, including customers, investors, and the broader community. Consistency between stated beliefs and actions enhances trust and credibility.

However, while beliefs are vitally important, they can also create challenges, especially in a changing environment. "Cognitive dissonance" is a psychological term that refers to the discomfort or tension experienced when a person holds two or more conflicting beliefs or attitudes.

Plato uses the cave story to illustrate the idea that people can become so attached to their beliefs and perceptions that they resist new information that conflicts with their existing worldview. The people in the cave experience cognitive dissonance when faced with new information that challenges their worldview, and they ultimately choose to reject it rather than accept the discomfort of reconciling their beliefs with new information. The allegory can be seen as a cautionary tale about the dangers of cognitive dissonance and the importance of remaining open to new ideas and perspectives.

Beliefs absolutely influence an organization's approach to innovation. Beliefs that encourage experimentation, risk-taking, and a growth mindset can foster a culture of innovation and help the organization adapt to changing environments and challenges. Or not.

In the late 1990s, I was devoting nearly all my time to assisting other retailers with developing their loyalty programs and helping them understand and strategically use shopper data. I was doing a lot of work with Brian Woolf at that time, whom some readers may recall as the "father" of loyalty marketing after the publication of his groundbreaking Coca-Cola research study.* Brian and I had been asked to meet with the management team of a regional retailer that was a subsidiary of a larger multi-banner retailer. The parent company had rolled out loyalty across their other banners and was aggressively using the data in how they went to market. This particular retail banner was a holdout.

It was the most frustrating day either of us had ever experienced. While the CEO and his team had been directed by the parent company to roll out loyalty, it was crystal clear that they didn't believe in it and were fighting it tooth and nail. They could not see a new possibility. Their beliefs—"loyalty doesn't work"—limited their view of the world and what was possible.

As Plato points out, understanding a different reality is hard. Skip ahead a couple thousand years, and Heather Heying and Bret Weinstein in their book *A Hunter-Gatherer's Guide to the 21st Century* write about the cognitive dissonance we as human beings experience

---

* Brian Woolf, *Measured Marketing: A Tool to Shape Food Store Strategy* (Self published, 1994), http://www.brianwoolf.com/index?action=bw_book_info&id=4.

as we move into a future increasingly foreign to us: The increasing rate of change in the modern world has outstripped the capacity of our brains and bodies to adapt. Humans have evolved in a local and linear world, knowing only the immediate environment and awakening each day secure in the knowledge that today will be a lot like yesterday and tomorrow will be a lot like today.

But that's no longer true.

Learning from the Past

The Industrial Revolution, spanning from the late eighteenth to the early nineteenth centuries, was a period of profound transformation that revolutionized the way goods were produced, distributed, and consumed. One of the key drivers was the invention and widespread adoption of new technologies, such as the steam engine, the power loom, and the spinning Jenny. These innovations dramatically increased the efficiency and scale of production, enabling businesses to produce goods at a previously unimaginable rate. The mechanization of manufacturing processes, combined with the utilization of new energy sources like coal and steam, fueled the growth of factories and led to the concentration of production in urban centers.

With the advent of factory production, the nature of work and employment underwent a profound transformation. The traditional artisanal system, where goods were crafted by skilled individuals in small workshops, gave way to a factory-based system characterized by the division of labor and specialized tasks. Workers became part of the labor force, working long hours in factory settings under strict supervision. This shift had significant implications for business owners, who now had to manage larger workforces and develop organizational structures to maximize efficiency and productivity.

Many businesses failed to evolve and adapt to the new industrial world; they believed their manual methods and processes were just fine and didn't see a need to change the way they had been doing things for many years. The companies that failed did so because they could not or would not change their beliefs; they could not acknowledge the changing world. The challenge was exacerbated by the rapid pace of technological change during that time, forcing businesses to constantly adapt and innovate.

Do these stories resonate as we move into the post-inflection point world, the world of Bionic Retail?™ Strategies, tactics, and practices that have worked for traditional retail in the past—like being a technology follower—no longer deliver success as we move into the future.

The mental disarray caused by waking up in the *Twilight Zone* of retailing can be severe. Nearly every day, we talk with retailers who do not understand how fast retail competition is changing, how decades-old business models are being disrupted, or how organizational cultures are being threatened. And more than anything else, we see retail leaders failing to understand that their worlds have been irrevocably changed as we speed into an unknown future.

Beliefs are important; they give form to an organization's culture. And the beliefs you and your organization hold, especially around new innovation, will determine your results as you move forward.

Results Don't Change until Beliefs Do

Ron Bonacci, vice president of marketing at Rouses Markets in Louisiana, is a true disciple of retail-loyalty marketing, having spent time at Kroger, KVAT Food Stores, United Supermarkets, and Weis Markets. He has helped each retailer drive their business forward while helping the owners and executive teams understand the power of shopper data.

But Bonacci is also a hard-nosed retailer. And like retailers everywhere, he inevitably looked to consumer packaged goods (CPG) brands and vendors for additional funding to support different loyalty initiatives. At the same time, he had the benefit of running and studying a multitude of different loyalty-based marketing initiatives, understanding what worked and what didn't. He's one of those unique left-brain and right-brain people who is comfortable with data while being incredibly creative.

Bonacci quickly understood that individual shoppers each have their own product brand preferences, sometimes being incredibly loyal to a brand and other times willing to switch amongst a limited number of acceptable brands. This understanding bumped up against brand-funded promotions that, logically, were financed by a CPG brand to attract customers to the brand. Except that's pretty hard to do; campaigns to promote a specific product or brand simply do not appeal to a good number of shoppers who prefer competing brands.

It was that combination, fueled by frustration with limited CPG funding, that drove him to try something different: he wanted to use all 40,000 SKUs in the store as the offer pool and provide savings on the products each customer wanted to buy, not what the CPG brands wanted to sell. Think about that for a moment: that means the retailer assumes funding for promotional discounts, not the vendors. Heresy in fast-moving consumer goods (FMCG) retailing.

Now, other loyalty executives saw and understood similar data. They could have come to the same conclusion as Bonacci and looked at things differently. But no one else did. They were limited by their beliefs. In this case, CPG brands and vendors have to pay for promotional discounts.

Bonacci secured the budget for his new initiative and pulled the trigger. And he knew it would work; he believed differently. And it did. Significantly higher numbers of shoppers redeemed the offers, providing them savings on the products they wanted to buy. The result: Bigger baskets, increased shopping frequency, and improved retention over time. And all at a very strong ROI. What more could a retailer ask for?

Your results in your business with new tech-fueled innovation won't change until you're able to consider new possibilities. And beliefs create the confidence to consider new possibilities.

Possibilities

I have had countless discussions with retail executives, including C-level executives at prominent companies, who speak very knowledgeably about innovation and seem to have a good grasp of industry transformation. But then, when discussion shifts to disrupting their own businesses through new technologies and practices, they quickly rationalize why those changes are not applicable to their company: "Well, that can't happen here."

Just as the Olympic skier must believe that winning the gold medal is possible before entering the starting chute, retail industry executives must believe that radical transformation is possible before embarking on the metamorphosis required to succeed in the post-inflection-point world.

Companies of any size across the massive retail industry must change their culture to embrace new possibilities and ongoing, ever-faster change. This challenge goes beyond business: People everywhere must prepare for a world of exponential change. And that's not easy to do.

Breaking out of our comfort zones to accept, if not embrace, change and new ways of doing things and becoming open to what's possible is a necessity. In my experience, people are capable of far more than they believe.

Some people are forced through circumstances to persevere—cancer patients having to battle through treatments, soldiers caught in life-or-death battles. Others knowingly seek to push themselves outside of their comfort zones by going skydiving or signing up for a marathon. My son Sterling has gone diving with sharks, ran marathons, and completed one of the toughest bike races in America, the Triple Bypass in Colorado, all to purposefully break out of his comfort zone. But all these challenges begin with a core belief that surviving the experience is possible.

Outward Bound is an organization that started in 1941 after Kurt Hahn, the founder, noticed that many sailors' lives were being lost at sea during World War II. The reason, he believed, was that these sailors lacked the mental fortitude to persevere in challenging circumstances. Thus was born Outward Bound as a school dedicated to physical fitness, enterprise, and physical and mental tenacity.

Each Outward Bound course is designed to provide a mental and physical challenge, be it hiking, canoeing, cycling, or even sailing. The last day is the culmination of the trip, a marathon-type event designed to prove to participants that they can indeed do more than they thought. I signed up for a weeklong cycling course in New England years ago, knowing the final day would be a 100-mile bike ride. The first days were educational but uneventful as I learned to rappel down cliffs, do ropes courses, and, of course, cycle up and down mountain passes carrying all my gear. All was going according to plan until, three days before the end of the trip, I took a bad fall off my bike, knocking myself out. After an emergency room visit, I was stitched up, bruised, and aching. Sitting out the next-to-last day—the program leaders were worried I might have a concussion—I thought about the final day, wondering if I should still try to ride. Having trained for months to prepare for the trip, I decided I had to at least try. I embraced the idea that completing the ride was possible.

Honestly, not knowing if I could ride even 1 mile, let alone 100 miles, I started out that morning with our group. Somehow, I finished, riding the 100 miles in about eight hours—with, it turns out, a couple of cracked ribs and a broken shoulder that were not picked up in the emergency room visit. Like I said earlier, we are all capable of far more than we give ourselves credit for when we commit to new possibilities.

Inspiration Can Strike at Any Time

In the early 2000s, I became fixated on the idea of true marketing personalization; I was frustrated by what I had learned about the inefficiency of mass promotion and the tremendous amount of money wasted by brands to create mass marketing campaigns that were not truly tailored to any specific customer. As mentioned earlier, Seventh Street Software had developed the first automated personalization engine but it was a standalone capability, it needed a surrounding system. In one of those classic middle-of-the-night epiphanies, I woke up at 2:00 a.m. to run to the kitchen and sketch out the picture in my head of the other pieces that had to be brought together, along with the data and workflow, to create what became called SmartShop, the first true personalized marketing solution for mass retail, in 2005. In this case, it was a relentless focus that eventually crystallized into inspiration. And that inspiration was to help the retail industry move from a world of mass promotion to a world of truly personalized, relevant savings.

Aligned Inspiration

A company's leadership and executive team can do things to prime their organizations for creativity and transformation. Maybe it's a dedicated couple of days offsite, maybe it's some physical adventure, or maybe it's turning the phones off and locking the office doors to get some time to focus on the future.

Whatever the forum, it's what happens that matters. And what needs to happen is for the leadership team to come into alignment; it is from that alignment that a successful culture is built. Aligned inspiration is the inflection point for organizations changing paths from the linear to the exponential. And that aligned inspiration creates commitment to a new path for creating the future of retail. That shared and aligned inspiration can change the world.

President John F. Kennedy, in a historic speech before a joint session of Congress on May 25, 1961, inspired the American people behind a grand vision to reassert American leadership in space exploration. “I believe that this nation should commit itself to achieving the goal, before this decade is out, of landing a man on the moon and returning him safely to the Earth.” This audacious goal set a clear and ambitious target for the nation’s space program and ignited a sense of national purpose and determination.

That vision and challenge aligned a vast nation to a singular vision: land a man on the moon within a decade. That aligned inspiration served to bring the government, the people, the military, NASA, and business together focused on a single goal.

I remember as a kid sitting in front of a small black-and-white TV on the afternoon of July 20, 1969, just over eight years after Kennedy’s iconic speech, as Neil Armstrong and Buzz Aldrin opened the hatch and stepped onto the moon. The moon landing represented a defining moment in human history and a testament to the power of bold leadership, shared inspiration, and collective effort. “That’s one small step for man, one giant leap for mankind.”

M&M Meat Shops: Inspired Vision to Committed Action

I’ll never forget the meeting I had with Mac Voisin, the founder and CEO of M&M Meat Shops, a unique retailer in Canada, some years ago. M&M had reached out to me because they were interested in developing and operating a loyalty program across their (at the time)

400 or so franchised stores. Making the trip to Kitchener, Ontario, I spent some time getting to know the team before launching into a presentation I had put together to help M&M understand what loyalty was all about and the power of it.

It was explaining a slide showing the significant differences in behavior and value of shoppers when I saw the light go on for Mac. He sat up straighter, became more intense, and spoke more strongly as he asked questions and dug into what had flipped the switch on. Gaining an understanding of true shopping behavior and customer value over time was an epiphany for Mac and his inspiration for committing himself and his entire organization to customer-intelligent retailing. And mind you, M&M was a franchise organization, with the different stores owned and managed by different people—a challenging environment in which to push a new initiative.

And commit he did.

While the marketing department took point, we spent significant time with other teams throughout the M&M headquarters organization, educating them around loyalty in general, the benefits we were focused on creating for M&M Meat Shops and the company's franchisees, and how the program would impact and benefit each of the departments within the company, such as Merchandising, Marketing, etc.

Mac launched the initiative at the company's bi-annual convention, where all the store owners got together. Mac asked me to prepare a presentation for the franchisees and the rest of the M&M team to help them understand the power of a loyalty program, the shopper data gathered through it, and how it could transform the franchisees' businesses. He backed that up by purchasing hundreds of copies of my first book to give to each of the store owners. It was this presentation that I had drafted my son Berkeley to dress up as a turkey for.

And, in a sign to the entire organization that he was serious, Mac "fired" a franchisee store owner who simply did not take the new initiative seriously and was producing poor results. That one act delivered a powerful message to all the other hundreds of store owners: Mac was serious. Mac had a vision. A new way for M&M Meat Shops to go to market and to operate. That inspiration was the inflection point for changing the path M&M was on.

M&M Meat Shops went on to roll out a national loyalty program, quickly achieving significant engagement with shoppers and integrating the understanding and use of shopper data into decision-making across their organization. And that success was triggered by one moment in time when the light bulb switched on in Mac's brain.

M&M's journey followed the blueprint that we've seen be successful time and again. Mac's inspired vision to use loyalty to build and solidify customer relationships at the store level and use the data gathered to improve decision-making across the organization was the keystone. Meetings with stakeholders across the company, both at headquarters and with store owners, surfaced the beliefs that could have become obstacles, enabling us to address them and set them aside. The vision turned into committed action as the program was pulled together, new technology was researched and secured to support the unique effort, the program was launched across Canada and implemented at store level. And then we embedded a focus on innovation and a new way of going to market into how the company operated.

That ongoing committed action was powerful. Twice a year the executive team of the company would travel to the different regions across Canada to meet with the franchisees in that area. Time was spent in each of those sessions to address any questions the store owners had, to brief them on the results and progress to date, share with them upcoming plans, and provide suggestions that they could take back to their stores. Those meetings, in addition to the period all-company conventions, served to embed and reinforce Mac's vision.

The Cost of Not Aligning Inspiration

The cost of not having aligned inspiration can be failure. Sometimes massive failure. In 2005, I sold the SmartShop technology and IP to Pay By Touch, a tech startup that had patents around the use of a biometric (finger scan) along with a PIN at checkout to access a digital wallet—basically Apple Pay way before its time. As the company raised increasing amounts of money, it gained momentum; many thought Pay By Touch would be the next unicorn, the next Microsoft or Google. And according to the buildup, it very well should have been.

The founder and CEO was fixated on creating a new payments company based on the use of biometric authentication at checkout. But the original vision—tying cheaper ACH payments to biometric authentication—failed to drive any meaningful customer adoption. Consumers were too invested in the rewards provided by the credit card companies to shift.

That realization led to hiring executives to explore the use of biometric authentication in a growing number of applications, from the casino industry to healthcare, cyber-security, and retail, effectively throwing spaghetti against the wall to see what would stick. Everyone involved "knew" that there was something big around biometrics, but the "what" had not yet come into focus.

Part of the acquisition of SmartShop involved my becoming the senior executive to lead the effort around tying personalized promotions to a finger scan, building on the success we had created in our retail store. As it became increasingly clear that providing personalized savings was the path to biometric adoption, the CEO and top executives—none of whom knew anything about retail—became increasingly involved, dictating sales approaches and business models, none of which aligned with retailers' needs.

What started out as a once-in-a-lifetime opportunity evolved into an incredibly frustrating experience. While everyone involved was excited about the potential of biometric authentication way before its time, the lack of alignment and lack of shared inspiration simply took all the energy out of the experience.

Pay By Touch raised nearly half a billion dollars by the beginning of 2007. But then, largely because of a failure to align inspiration, everything fell apart. As the economic crisis of 2007–2008 hit, cash dried up, and Pay By Touch entered bankruptcy later that year.

For more on changing beliefs, opening up to new possibilities, and inspiration, I highly recommend my son Sterling's book, *Hunting Discomfort*. In it he leverages incredible personal experiences, along with years of research, to help people and organizations achieve more . . . no matter what. He shares what he has learned with audiences around the world as a top-rated speaker. Learn more at sterlinghawkins.com

Chapter Summary

- The most important part of transforming yourself and your organization for this new world is not technology, it is altering your perceptions and beliefs.
- Organizations have collective beliefs that create a company's culture. When people in an organization share common beliefs, it promotes unity, alignment, a common purpose, and collaboration.
- Your results in your business with new tech-fueled innovation won't change until you're able to consider new possibilities. And beliefs can either limit your possibilities or create the confidence to consider new possibilities.

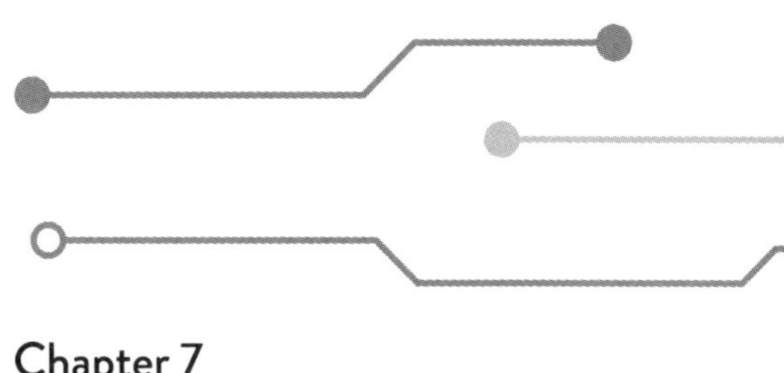

Chapter 7

AN INNOVATION PHILOSOPHY

"I operate on the physics approach to analysis. You boil things down to the first principles or fundamental truths in a particular area and then you reason up from there."

—Elon Musk

Elon Musk's SpaceX applied first-principles thinking to space exploration, specifically the high costs associated with rocket launches. Instead of accepting the prevailing industry norms, Musk broke down the problem into fundamental principles and sought to develop reusable rockets. By challenging the assumption that rockets should be discarded after each launch, SpaceX successfully designed, tested, and implemented reusable rocket technology, significantly reducing the cost of space missions.

Changing beliefs, opening up to new possibilities, and finding and aligning inspiration are all necessary to support leaping onto the exponential path to retail's future. But they alone are not enough. We need something more specific and concrete to move us forward in driving innovation and recreating our businesses. And who better to turn to than perhaps the world's greatest entrepreneur, Elon Musk.

Musk was born and raised in South Africa, eventually moving to the United States, where he received a degree in physics from the University of Pennsylvania and a degree in economics from the Wharton School. His original startup, Zip2, was founded in 1995 with $15,000 and was acquired by Compaq Computer for $341 million four years later. Using the buyout money, he created X.com, an early fintech company that was later merged into PayPal. Musk then parlayed the proceeds from PayPal's acquisition by eBay into funding SpaceX, Tesla, and SolarCity.

Musk has gone on to found several other notable companies, including The Boring Company, Neuralink, and X Corp., and fund numerous others, such as his acquisition of Twitter. And then there are the new innovations coming out of his existing companies, like Optimus, the humanoid robot that Tesla is developing. Today, Musk is the world's richest person, sometimes trading the number one and two spots with Bernard Arnaud, the CEO of luxury goods company LMVH.

Through this journey, his education in physics has remained a cornerstone of his management style: "Physics is a good framework for thinking," he would say later. "Boil things down to their fundamental truths and reason up from there."*

Fundamental truths are a foundational element of what is known as first-principles thinking. First-principles thinking is a problem-solving approach that involves breaking down complex problems or situations into their fundamental truths or basic principles and then reasoning from there. It is a method of thinking that encourages individuals to question assumptions and traditional wisdom in order to gain a deeper understanding and uncover new insights. First-principles thinking usually rejects herd-mentality theories to discover improved or completely new paths to solutions. Just the thing to create the future of retail.

An Innovation Philosophy

The concept of first-principles thinking has its origins in ancient Greek philosophy, particularly in the works of pre-Socratic philosophers, and later developed by Aristotle. These philosophers sought to understand the fundamental principles underlying the natural world and human existence, and their inquiries laid the groundwork for rational and systematic thinking.

It was Aristotle who provided a more systematic and comprehensive understanding of first-principles thinking. Aristotle believed

---

* Tom Popomaronis "Elon Musk Calls This a 'Powerful, Powerful Way of Thinking'—But Is 'Hard to Do.' Here's How It Works," February 29, 2020, https://www.cnbc.com/2020/02/28/billionaire-elon-musk-this-is-a-powerful-way-of-thinking-but-hard-to-do-how-it-works.html#:~:text=In%20a%202013%20TED%20talk,opposed%20to%20reasoning%20by%20analogy.%E2%80%9D

that knowledge could be derived from observing and analyzing the world around us. He distinguished between two types of knowledge: *episteme* (knowledge of universal principles) and *techne* (knowledge of specific techniques or crafts). The term "technology" comes from the Greek *techne* and the word *logos*, meaning word.

Aristotle emphasized the importance of starting with first principles, or *archai*, as the foundation for knowledge. He argued that first principles were self-evident truths that did not require further proof and that all subsequent knowledge could be derived from them through logical reasoning. Aristotle's approach to first-principles thinking formed the basis of his philosophical system and scientific methodology.

A key aspect of Aristotle's philosophy was his concept of the Four Causes, which aimed to understand the nature and explanation of things. These causes were:

- The material cause: the substance from which something is made
- The formal cause: the defining characteristics or structure
- The efficient cause: the agent or force that brings something into existence
- The final cause: the purpose or goal of something

By understanding these causes, Aristotle believed a person could gain a deeper understanding of the essence and nature of phenomena.

First-Principles Thinking

Long-held beliefs and a "we've always done it this way" mentality too frequently prevent innovation in retail. Many times, the perpetuation of long-accepted practices in the retail industry is due to executives who have spent their career with a given company or in a particular retail channel without challenging themselves to understand other perspectives or approaches. While working their way up through the ranks resulted in respect from peers, the thought processes and assumptions gained along the way create limitations in a world where, increasingly, anything is possible.

Breaking down those established assumptions can mean the difference between incremental improvement and exponential growth. Too often, retailers simply adopt and deploy solutions that have been developed by some third party to address a specific problem or opportunity rather than take a fresh approach. Think of it like buying a suit off the rack vs. a bespoke suit fully customized to you. Rather than accepting conventional wisdom or existing solutions that have generalized characteristics to accommodate varying organizations, first-principles thinking provides a proven process for exploring new possibilities that could solve an organization's unique problems.

In the context of retail innovation, first-principles thinking prompts retailers to start fresh in exploring how retail operates, customer shopping behavior, and how value is created. By approaching retail with a blank slate, retailers can reimagine the supply chain, store operations, product and service offerings, and the entire customer journey.

First-principles thinking typically follows a process:

1. **Identify and define the problem:** Clearly articulate the problem or challenge you are trying to address.

2. **Break it down into fundamental principles:** Break down the problem into its essential components and identify the underlying principles or facts that are known to be true.
3. **Challenge existing assumptions:** Question any assumptions or conventional wisdom that may be taken for granted. Ask yourself why things are the way they are and whether there are alternative approaches.
4. **Build from the ground up:** Reconstruct the problem or situation using the fundamental principles identified in step 2. Avoid relying on analogies or existing solutions and instead start with a clean slate.
5. **Reason from first principles:** Use logical reasoning to derive new conclusions or solutions based on the fundamental principles. This may involve exploring different possibilities, considering different perspectives, and thinking creatively. This is where changing beliefs and new possibilities come in.
6. **Test and refine:** Once you have derived new conclusions or solutions, test them against real-world constraints and gather feedback. Refine your thinking and iterate as necessary.

First-Principles Thinking in Action

Thinking back, when I created SmartShop twenty years ago, I unknowingly followed the first principles process.

1. Identify and define the problem.

Customer purchase data gathered through our loyalty program brought to life true shopper behavior. While we had a handful of extremely loyal shoppers who did nearly all their grocery shopping with us, the large majority of shoppers were obviously doing some shopping with us *and* with other retailers. Digging deeper, we observed many shoppers pursuing the weekly ad deals, going store to store to save money. So, the problem with growing weekly sales was finding a way to get more (occasional) shoppers into the store on a more regular basis. CPG brand-driven weekly ads created the opposite of loyalty; they incentivized shopper promiscuity.

2. Break it down into fundamental principles.

So, the essential components of the problem were:

1. Savings are important to nearly all shoppers.
2. Shoppers have brands and products they are loyal to and will go to different retailers to find savings on those products.
3. CPG brand promotional funds drove weekly ads, which caused shoppers to go store to store, exacerbating the core problem.

3. Challenge existing assumptions.

As we brainstormed through different ideas to address the problem—how to get more shoppers in the store more often—we ran through the typical ideas, including putting more products on sale each week, moving to an every day low price (EDLP) program, and others. We re-reviewed the data to confirm the shopping behavior that we were focused on. We discussed other reasons shoppers may visit other stores, for example, the availability of other brands and products, pricing, services, etc.

We also thoroughly discussed how grocery retail traditionally went to market, driven by CPG brand marketing funds and promotions. Retailers rely on CPG brands to fund most, if not all, of the discounts provided to shoppers each week in the weekly ad, TPRs (temporary price reductions), and other deals.

4. Build from the ground up.

So, our practice of distributing weekly ads filled with mass promotions was actually encouraging disloyalty; we were actually driving shoppers to visit other stores to maximize the value for their money. Further, we saw in our data that each shopper has their own brand preferences and discount propensities. Mass promotion was causing the challenge and marketing relevancy was the answer.

5. Reason from first principles.

It was only because of my growing knowledge and comfort with technology that we could begin with a clean sheet of paper to design a new approach. The goal was to create a capability that provided savings on products important to each shopper, which we believed would result in stronger shopper loyalty. The SmartShop system ultimately included the ability to create an offer pool to use in targeting, the ability to establish "guardrails" (minimum and maximum discounts, quantities, etc.), the ability to optimize the outcome to a given budget, the ability to communicate the personalized promotions to shoppers, and then deliver the personalized discounts at checkout.

6. Test and refine.

Upon launching SmartShop, we iterated through several learnings, including the impact of a limited offer pool vs. using the entire store product catalog, the importance of the purchase cadence of individual products, favored communication channels, and more.

Looking back, the hardest part of developing, launching, and operating SmartShop was a cultural problem: getting our management team to wrap their heads around the idea of cutting back on the traditional weekly ad and relying on new technology to go to market differently, providing personalized savings to each shopper. It was having actual data, gained from our loyalty program, that provided the factual information for understanding how customers were truly behaving and gave us the ability to break down long-held assumptions.

Driving Retail Innovation & Creating the Future of Retail

So, it's important to start thinking about how you could use first-principles thinking to drive innovation in retail. To understand this approach and how it affects several facets of retail management, we'll discuss what first-principles thinking could look like in a few major aspects of the retail industry.

Rethinking the Customer Experience

The customer experience is the most important part of running a successful retail operation. Using first-principles thinking means looking at the basic ideas of customer satisfaction, convenience, and delight. Retailers can start by questioning common ideas about how customers use goods, buy things, and interact with brands. Many retail executives believe all shoppers are motivated by low prices and savings.

For example, think about using tech like augmented reality to educate and inform shoppers about the products they're considering. By rethinking the customer experience from the ground up, businesses can open up new ways to connect with customers and stand out in the market.

Also consider the myriad ways to improve the actual checkout and purchase experience. Tech like computer vision, smart carts, and even self-shopping apps can save the usual standing in line waiting to checkout. Biometric payment capabilities can speed up the payment process, eliminating potential friction.

Rethinking Store Design and Plan

Most stores in FMCG retail are designed for efficiency and deliberately moving the shopper around to higher margin departments and categories; retailers that approach design from a shopper experience or retail-as-theater philosophy are few and far between. I believe there is an opportunity to apply first-principles thinking to design and create new concepts in how customers shop a store and discover new products and how services are delivered.

Also, thinking from first principles can challenge traditional ideas about how to show and arrange products. Retailers can try out new ways to show off their goods by using technologies like smart shelves, interactive displays, and dynamic pricing systems. By starting with the basics, retailers can create immersive and interesting settings that draw in customers and make shopping a better experience overall.

Rethinking Retailer–CPG Brand Manufacturer Relationships

Fast-moving consumer goods retail—think supermarkets, convenience stores, drug stores, discount stores—is very reliant on CPG brand manufacturers, and not just for products. CPG brand manufactures collectively spend over $200 billion annually on marketing, with a significant part of those funds going to retailers in support of trade promotion, shopper marketing activities, and, increasingly, retail media networks (RMNs).

Those trade funds can mean the difference between profit and loss for many retailers as some portion of those monies contribute to category gross margins. Those funds have caused many retailers to become product-driven rather than customer-focused. First-principles thinking is an opportune time to dissect and reinvent that relationship.

Supply Chain and Logistics Innovation

The supply chain and logistics are very important parts of the shopping business. Using first-principles thinking in these areas means looking at the basic ideas behind efficiency, cost-effectiveness, and sustainability.

Retailers can start by asking if standard supply chain models are the best way to do things and looking into other options. For example, instead of depending solely on a centralized distribution system, retailers can start from first principles and think about decentralized fulfillment centers or micro-fulfillment solutions that bring products closer to the end consumer.

Another element to consider in supply chains is growing customization and personalization of products, not just the marketing and pricing of products. In the food industry, as a growing number of consumers have specific dietary requirements, there is opportunity to customize prepared foods to the specific shopper.

Technology and First-Principles Thinking

First-principles thinking is an important tool as companies follow the blueprint to creating their future. Moving from an inspired vision to committed action, first-principles thinking is a proven way to identify and then move past beliefs and practices that need to transform on the journey into the future. And that future is built on tech-first thinking.

Chapter Summary

- First-principles thinking prompts retailers to start fresh in exploring how retail operates, customer shopping behavior, and how value is created.
- Retail leaders can leverage the six steps of first-principles thinking to foster innovation: identify and define the problem; break it down into fundamental principles; challenge existing assumptions; build from the ground up; reason from first principles; and test and refine.
- First-principles thinking can catalyze growth in any area of the retail industry, including customer experience, store design and layout, and logistics, among many others.

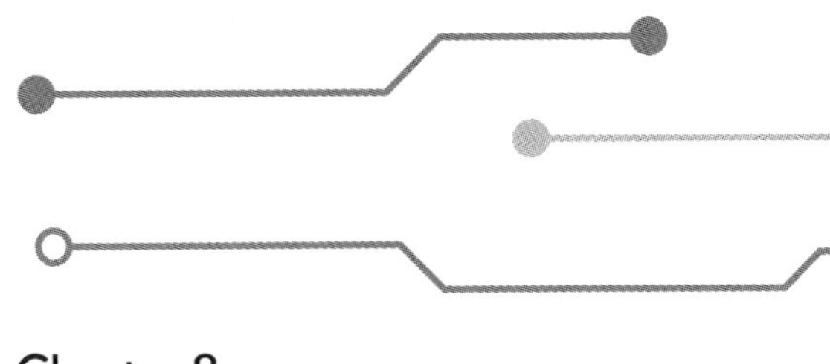

Chapter 8

TECH-FIRST STRATEGY

"We often talk about tech companies, how important tech companies have been to the U.S. economy . . . But the truth is we need to be tech companies. In my view, the battle that we're all about to face is how can we all become great tech companies before tech companies become greater retailers? That's the fundamental challenge that we face."

—Greg Boyan, CEO of HEB, speaking at the FMI Midwinter Executive Conference 2023

We've all heard the parable of the boiling frog. The idea that if a frog is put into a pot of boiling water, it will instantly jump out. But if a frog is put into a pot of lukewarm water and the pot is slowly brought to a boil over time, the frog will not sense any danger and will be cooked to death.

I think there's an analogy here to what's happening as retail leaders have become increasingly immersed in technology-driven innovation over the past decade. Retailers are being slowly pulled along with technology's evolution without feeling any impending doom. And yet, as we've passed the inflection point, the water is beginning to boil.

Battle to Create the Future of Retail

What is clear is this: We have surpassed the tipping point on Kurzweil's exponential growth curve of technology, and technology is already beginning to outpace and outthink human beings. Stop to think about that. Billions of people alive today, many retail executives included, will experience firsthand a world where AI systems are faster and smarter than their human creators. It's already happening in retail, where AI-powered expert marketing systems are able to digest vast amounts of data on each individual shopper, develop a unique strategy for each shopper to grow value and engagement, and generate targeted promotions and pricing to achieve each strategy, all in an automated process that takes minutes for millions of shoppers. No human being, or army of human beings, can do this.

The pace of change, especially technological change, outside the organization is growing (much) faster than the pace of change within retail organizations. And as that gap grows ever-larger, traditional retailers will feel increasing pain.

As we have surpassed the inflection point on the path to retail's future, the industry has entered a new battlefield for which it is poorly prepared. What most retailers are missing is that there has been a fundamental shift in the world, that technology now drives business strategy, and that digital-native companies with a tech-first mentality are out to reinvent brick-and-mortar retail. And that companies like Amazon, Walmart, and Kroger are much further up and out of the technology growth curve than almost everyone else.

The shift in perspective is this: Traditional retailers look at what's happening and view it as technology impacting retail. I look at what's happening and believe that retail is metamorphosing into a technology business. A technology business that just so happens to sell things, both online and in brick-and-mortar stores.

And I'll go even further: I believe that there is a battle underway to create the future of retail—not just incrementally change it a step at a time, but fundamentally change the way shopping is experienced. And I believe that a handful of retailers, companies like Amazon, Walmart, and Kroger, have joined that battle and are considerably ahead of their competitors in creating retail's future.

Key leaders have a vision for what they feel the future of retail looks like for their organizations. Companies like those above have accepted that nearly anything is possible today and are leveraging technology to realize their visions.

Just look at what Walmart is doing. "Two years ago, President and CEO Doug McMillon described the company's future business model as being based on three main pillars: e-commerce, financial services, and health and wellness. But at Consumer Electronics Show (CES) 2023, Walmart executives detailed how that model has evolved to comprise a larger strategic innovation framework with three additional pillars: media, advertising, and fulfillment services."* To see how Walmart is bringing this expanded flywheel to life, look no further than the company's partnership with Salesforce

---

* Gina Acosta, "Walmart Doubles Down on New Flywheel Business Model," Progressive Grocer, April 17, 2023, https://progressivegrocer.com/walmart-doubles-down-new-flywheel-business-model.

to make Walmart's last-mile fulfillment and delivery service available to other retailers.

Walmart's focus on their marketplace, advertising platform (retail media network [RMN]), data analytics, and myriad other technologies show just how committed they are to transforming beyond a traditional retailer to a technology-retail hybrid, a bionic retailer.

More than Digital Transformation

Coming off pandemic-driven sales increases and inflation-fueled revenue and margin gains, retailers are riding high. Traditional retailers believe they are in good shape and point to the investments they are making in new technology and innovation. According to FMI (Food Industry Association) research, supermarket retailers in 2022 invested more than $13 billion* in technology and are projected to increase those investments in subsequent years.

That $13 billion technology investment sounds good. But, based on my subjective thoughts resulting from speaking with many retailers across the industry, much of that spend continues to go to replacing or rebuilding core systems that should have been updated years ago. That and a good portion of that spend going to license fees, cloud fees, etc., not new innovation. And compare that $13 billion spend across many retailers to Amazon, who invested over

---

* Catherine Douglas Moran, "Food Retailers Spent More than $13B on Tech Investments in 2022: FMI," Grocery Dive, July 13, 2023, https://www.grocerydive.com/news/grocers-invested-13-billion-in-technology-2022/686872/?utm_source=Sailthru&utm_medium=email&utm_campaign=Issue: 2023-07-13 Grocery Dive [issue:52402]&utm_term=Grocery Dive.

$72 billion[*] in tech research and development the same year, 31 percent more than the year prior.

And Amazon is not the only one making major investments. "In 2019, Walmart invested $11 billion in technology, an increase of 40 percent from 2018. This investment allowed the company to open new e-commerce fulfillment centers, develop new mobile applications for customers, and expand its online grocery delivery services. In 2020, Walmart reported having more than 1 million robots working across its stores and warehouses worldwide. Furthermore, the company has also implemented AI-powered chatbots on its website and mobile app to give customers faster responses to their queries."[†]

A few years ago, the MIT Sloan Review released a report on digital transformation in collaboration with Deloitte that I think provides a helpful framework for thinking about the transformation traditional retail is undergoing.

The authors explain that companies early in the transformation process focus on using technology to solve discrete business problems. Many retailers are at this stage, looking to secure and deploy tech capabilities to solve specific operational problems. We saw this in the early stages of the pandemic as retailers without any online shopping capability ran to Instacart or hurriedly put in place other eCommerce platforms.

---

[*] MacroTrends, "Amazon Research and Development Expenses 2010-2023 | AMZN," accessed September 2023, https://www.macrotrends.net/stocks/charts/AMZN/amazon/research-development-expenses.

[†] Bradley Pallister, "How Does Walmart Foster a Culture of Innovation," Innovolo Group, March 13, 2023, https://innovolo-group.com/innovation-en/innovation-insights-en/how-does-walmart-foster-a-culture-of-innovation/.

As companies evolve through the digital transformation, they begin to focus more on integrating the disparate capabilities they have deployed across their organizations, seeking efficiencies and better use of the technologies they've already invested in.

We see examples of this in the market, often among the larger regional retailers. For example, retailers can integrate computer-assisted ordering and virtual inventory systems with eCommerce platforms to automatically withdraw out-of-stock items from the digital storefront. Or KVAT Food City integrating Sifter's shop-by-diet filters into its eCommerce site, or Strack & Van Til integrating Birdzi's sophisticated marketing personalization into their digital stack.

I agree with both of these transformation phases: an initial focus on solving discrete problems, then evolving to integrating capabilities to realize efficiencies and an improved shopper experience. But it's in the next phase that I diverge from MIT Sloan's report. The report suggests that more digitally mature businesses use technology to support and execute the company's strategy. I am of the opinion that technology itself opens the door to new business strategies.

What Comes First, Technology or Strategy?

Jeff Bezos launched Amazon in 1994 after working in the financial services industry and seeing the potential of the internet. Bezos decided Amazon's initial foray into the nascent world of online commerce would be books, seeing the large demand for literature and the huge number of titles available. "I picked books because there were more items in the book category than in any other category. And so, you could build a universal selection. In 1994, when I was pulling this idea together, there were 3 million different books active in print at any given time. The largest physical bookstores only had about 150,000 different titles. And so, I could see how you could make a bookstore online with a universal selection. Every book ever printed, even the out-of-print ones, was the original vision for the company. So that's why books."*

What Bezos did not do was first decide that he wanted to sell books and then look for the best way to accomplish that. Bezos saw the enormous potential of the internet (technology) and how it could change commerce. After deciding that he wanted to participate in the growth of the web, he explored what products to initially sell (books). So, in a very real sense, it was a technology-driven strategy.

Contrast the Amazon story with the invention of self-checkout. When self-checkouts were introduced in 1986, retailers were looking to lower labor costs and improve efficiencies. Strategy (lower costs) was enabled by technology (self-checkout).

The question of what comes first, strategy or technology, is the proverbial chicken-and-egg dilemma. And I would suggest that there is no one answer that applies all the time. History is replete with examples of companies using, and sometimes creating, technology to enable business strategy.

But I will also suggest that as we move up and out of the exponential growth curve of technology, insight into new innovation, understanding how different technologies work, and creativity in how technologies can be applied and how they converge will

---

* Michael Kozlowski, "Amazon CEO Jeff Bezos Talks about Selling Books in New Interview," May 9, 2028, https://goodereader.com/blog/business-news/amazon-ceo-jeff-bezos-talks-about-selling-books-in-new-interview.

increasingly come together to create and enable new business strategies in the retail industry.

Using technology to facilitate business strategy—like self-checkout—has become almost a part of standard operating procedure at larger retailers as they undergo digital transformation. Seeing in technological innovation the opportunity for new business strategies remains something foreign to many traditional retailers. But digital-native companies are born thinking tech-first and represent what I believe will be a significant challenge to traditional retailers in the time ahead.

Tech-First

In his oft-cited 2003 *Harvard Business Review* article, "IT Doesn't Matter," Nicholas Carr argued that unless a technology is proprietary to a company, it ultimately won't provide a competitive advantage on its own. As was the case with electricity and rail transport, many technologies will become available to all and thus provide no inherent advantage. The trap to avoid, according to Carr, is focusing on technology as an end in itself. Instead, technology should be a means to strategically potent ends."*

Carr wrote that article twenty years ago, not knowing or understanding the technological developments that would occur over the

---

* Gerald C. Kane, Doug Palmer, Anh Nguyen Phillips, David Kiron, and Natasha Buckley, "Strategy, Not Technology, Drives Digital Transformation," *MITSloan Management Review*, Research Report, 2015, https://www2.deloitte.com/content/dam/Deloitte/fr/Documents/strategy/dup_strategy-not-technology-drives-digital-transformation.pdf.

past two decades. And I think it's that gap that led him, like the MIT Sloan report, to suggest that technology should be used to support and facilitate strategy. Seeing what has transpired in the last twenty years, how technology has developed, how digital-native companies think and operate, tech startup behavior, and more, reinforces my view that technology innovation can and should increasingly drive company strategy.

This is admittedly a challenging change to make for retail leaders. Given that technology is increasingly able to make most anything possible, I think the place to start is with a blank sheet of paper. Think about the problem you're trying to solve, or what you believe the opportunity is, and consider how you would accomplish it without the preconceived notions of past practices or beliefs. Not easy to do but does become easier with practice. Thinking tech-first may also be easier as part of a team discussion, where others can reinforce the change in perspective, keeping each other focused on creating some new product, process, or service, using technology.

While electric cars have actually been around for more than a century, it was not until Musk brought innovative technology to the market that electric vehicles became a mass market. Tesla's development of innovative electric vehicle technology, including high-performance batteries and self-driving capabilities, has allowed the company to disrupt the automotive industry and establish itself as a leader in the electric car market.

Apple's launch of the iPhone in 2007 revolutionized the mobile phone industry and transformed Apple's business strategy. The iPhone's innovative technology, including its touch screen and user-friendly interface, enabled Apple to capture a significant share of the smartphone market and establish itself as a leading technology company.

The iPhone opened the door to creating Apple's App Store, which eventually became a major source of revenue for the company. Apple didn't think, *Gee, let's create a store where people can download apps. Oh, and I guess we'll need a device to download the apps to.* The technology of the iPhone and the App Store launched a major new business that has transformed how consumers access and use technology.

Amazon's development of Amazon Web Services (AWS) in 2006, which provides cloud computing and storage services to businesses, enabled the company to diversify its revenue streams and establish a dominant position in the cloud computing industry. According to Tom Furphy, who led the team that developed Amazon's Subscribe & Save replenishment product, AWS was originally built as a set of services and server capacity to enable Amazon teams to build out their tech capabilities and capacity ahead of demand. By building ahead, being the first adopter of their own services, and then leasing out the excess capacity, Amazon effectively turned a cost center into a revenue-generating business.

Perhaps learning from Amazon, Walmart has partnered with Salesforce to provide delivery services for other retails via the Salesforce AppExchange. Extending Walmart's GoLocal service to other merchants, enables the company to build up its own delivery platform while reducing reliance on other third-party services like DoorDash and Uber.

Tech-First Thinking

Technology-first thinking refers to an approach or mindset where technology is considered the primary driver or enabler of innovation, problem-solving, and business strategy. It involves placing technology at the forefront of decision-making processes and considering its potential and capabilities before other factors.

In a technology-first approach, organizations prioritize the exploration and adoption of new technologies as a means to achieve their business objectives. They actively seek out technological advancements, emerging trends, and disruptive innovations to gain a competitive advantage and create value for customers. Tech-first companies actively create their future.

A mindset that prioritizes technology has several defining features:

- Businesses that put technology first are always on the lookout for innovative solutions and are willing to try something new. They foster an **aggressive adoption of technology**. They are always looking out for new innovations in technology and investigating how they may improve their business and the way they serve their customers.
- To solve business problems and meet client demands, technology-first thinkers **prioritize technological solutions.** Their goal is to find cutting-edge technology that will help them streamline operations, cut costs, and create ground-breaking new goods and services.
- A technology-first mentality fosters a culture of innovation by challenging everyone in the organization to **envision new ways in which technology can improve the customer**

experience or operations. This culture change supports trying new things, taking calculated risks, and gaining insight from setbacks.

- Tech-fueled innovation is **largely focused on the shopper**, improving the shopper experience and creating new products and services.
- Organizations that adopt technology-first thinking are **dedicated to constant improvement.** They devote resources and create processes to monitor the market for new developments and changing shopper expectations.
- Working with others is essential to a technology-first mindset, whether that's established businesses, new companies, or even individual experts. Companies understand that in a fast-changing world it's impossible to always have the requisite skill sets and knowledge; **partnerships with other businesses, experts, and colleges and universities prove invaluable.** Kroger values partnerships: "University partnerships are some of Whitacre's favourite interactions in the Kroger Labs ecosystem. The lab team works with multiple universities and colleges to answer unique questions, working directly with faculty and students, and even has physical space on both the University of Cincinnati and Northern Kentucky University campuses—each located minutes from the company's headquarters."*

* Kroger, "Redefining the Grocery Customer Experience," March 21, 2022, https://technologymagazine.com/company-reports/kroger-redefining-grocery-customers-experience.

Dual-Paths

Unless a technological capability is truly proprietary, capabilities eventually become commoditized, and this happens ever faster in the exponential world. This means that retailers must be thinking along two paths: maintaining a competitive stance relative to competitors while innovating to differentiate. And always be looking to see where you can create proprietary advantage.

New technologies are enabling more significant shifts in costs, shopping experiences, and marketing efficacy. As has always been the case, retailers need to respond to any significant competitive technology deployments that change the status quo in a meaningful way. A textbook example of that is automation.

In the summer of 2021, Walmart announced it would be deploying Symbotic's automation to twenty-five of its distribution centers. In May 2022, Walmart expanded the relationship, announcing it would install Symbotic's automation platform in all of its forty-two distribution centers. That announcement was followed by Walmart taking a significant stake in Symbotic,* seeing the technology as vital to its future, similar to Amazon acquiring Kiva Systems robotics around a decade ago.

And here's the kicker: Walmart's inventory sits within 10 miles of 90 percent of the U.S. population. By the end of 2026, 55 percent of its fulfillment distribution center volume will move through automated facilities. The company claims that unit cost averages will fall

* Christine LaFave Grace, "Walmart Takes Majority Stake in Symbotic," Winsight Grocery Business, June 21, 2022, https://www.winsightgrocerybusiness.com/walmart/walmart-takes-majority-stake-symbotic.

by about 20 percent.*

Talk of efficiency and cost savings through automation is one thing. Now we have the world's largest retailer claiming it is able to reduce unit cost averages by some 20 percent. In the razor-thin profit margins of fast-moving consumer goods retail, that's a game-changer, making the automation "butterfly" disruptive. And we are already seeing competitive responses.

Target and Albertsons were already customers of Symbotic before Walmart's involvement, but just recently Associated Food Stores, out of Salt Lake City, Utah, signed on with Symbotic to automate their distribution center as AFS seeks to maintain distribution cost parity with Walmart. Others, like Associated Wholesale Grocers, are deploying automation using a different solution provider.

So, distribution center automation is significantly altering operating overhead cost, upsetting industry equilibrium, and forcing competing retailers to react so as to maintain cost parity. Cost parity is one thing, but retailers also need to consider maintaining parity around the shopping experience and digital user experience.

Walmart, Target, Amazon, Kroger, and Albertsons understand the importance of digital user experience; sadly, many other retailers do not yet appreciate how important this is. Far too many retailers have cobbled together their digital infrastructure from multiple solution providers over time and then try to integrate the disparate pieces. Inevitably, this results in the opposite of the seamless, cohesive, and comprehensive user experience shoppers are coming to expect.

---

* Daphne Howland, "Walmart Banks on Stores, Robots as It Stokes e-Commerce," April 5, 2023, https://www.grocerydive.com/news/walmart-automated-stores-robots-e-commerce/646918/.

Maintaining competitive position is one path, innovating around shopper services and engagement is a parallel path.

Amazon's new Amazon One biometric payment solution is an example of innovating to provide differentiated customer services. The technology is based on touchless biometric "scanning" of your palm to access your Amazon payment wallet. When it was first installed at our local Whole Foods, I was able to enroll my palm in just seconds. Amazon knew who I was through scanning my Whole Foods QR code that provides Prime savings. It then created the link between the credit card stored in my Amazon wallet and my palm, enabling me to enroll in less than a minute. This is differentiated innovation.

Amazon's Alexa platform is also positioned to provide innovative differentiation. Voice technology is poised to explode, driven by advances in processing power and artificial intelligence. Amazon is well-positioned to leverage new voice capabilities, taking advantage of the Alexa platform that extends from the home to the car, and from the office to the store. Alexa stations positioned throughout Amazon's Fresh format stores already provides shoppers with assistance in locating products, meal and recipe suggestions, and more. Again, this is an example of differentiated innovation.

Implementing the Future

The past few chapters have been devoted to helping retailers understand the need to change beliefs and open up to the new possibilities afforded by technological advancement. First-principles thinking provides a philosophy and tech-first thinking a strategy for moving forward.

What's next is embedding this learning into the organization's infrastructure. And retailers can do that by developing an innovation doctrine.

Chapter Summary

- Retail is metamorphosing into a technology business. This mindset shift is imperative for traditional retailers to pivot into the tech-first mentality necessary to compete in a world past the inflection point.
- In a technology-first approach, organizations prioritize the exploration and adoption of new technologies as a means to achieve their business objectives. They actively seek out technological advancements, emerging trends, and disruptive innovations to gain a competitive advantage and create value for customers.
- Due to the rate of change and technology adoption, retailers need to be focusing on two paths simultaneously: maintaining a competitive stance relative to competitors and always searching to identify where you can create proprietary advantage.

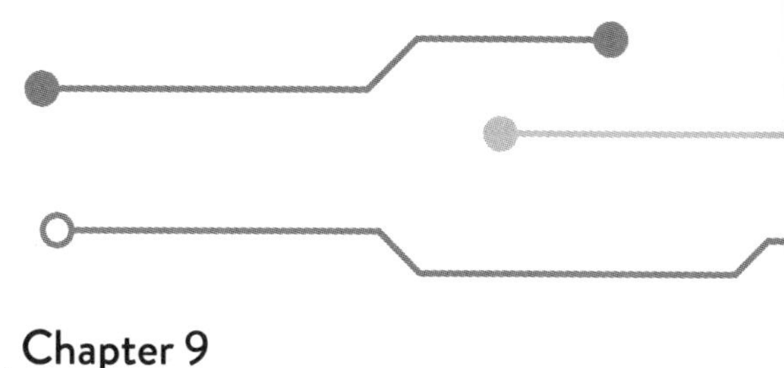

Chapter 9

AN INNOVATION DOCTRINE

Turning on CNN the evening of January 17, 1991, we watched the opening salvo in Operation Desert Storm, the Gulf War, unfold in real time. The aerial bombardment of strategic military and infrastructure sites in Iraq, and especially Baghdad, was just unbelievable to watch as Peter Arnott and Bernard Shaw provided ongoing commentary from inside Iraq.

Desert Storm was a textbook example of U.S. military doctrine in action, focused on joint operations, massive firepower, and rapid maneuverability. The air assault that began that day was designed to destroy critical defense and communications infrastructure. As the U.S. Air Force degraded Iraq's capabilities, the ground offensive kicked off, leveraging the ability to rapidly deploy and sustain forces over a long distance. The overwhelming force the U.S. brought to the conflict made for a quick victory, the battle lasting only about forty days.

Military doctrine refers to the set of principles, strategies, and guidelines that govern the organization, planning, and conduct of military operations. It serves as a framework for how a military force envisions and prepares for battle, and it provides a common understanding and approach for commanders and personnel at all levels. It is a guide to action, rather than hard-and-fast rules. It provides groups working together with a common purpose and direction.

We believe that retailers can employ a similar methodology, creating an innovation doctrine to guide the retailer's organization into the future. We can use the wide-ranging elements of military doctrine as a framework for developing a doctrine appropriate to retail innovation.

Military Doctrine / Strategic Concepts: *These are overarching principles that guide the military's approach to achieving its objectives. They may include ideas such as deterrence, defense, offense, and escalation management.*

Retail Innovation Doctrine and Strategic Concepts

Culture of Innovation

A culture of innovation involves encouraging employees to embrace a growth mindset, take risks, and think creatively. Organizations can promote innovation by providing resources and tools for experimentation, establishing cross-functional teams, and encouraging collaboration. And if there was one thing within an organizational culture requisite for success it is fostering innovation. "According to a survey by McKinsey, 87 percent of executives believe that innovation is key to their company's success. Additionally, research from the *Harvard Business Review* found that companies with an innovative culture are five times more likely to be successful than those without one."*

Failing fast is an important part of building a culture that embraces innovation, experimentation, and continuous improvement. When retailers create an environment that encourages risk-taking and learning from failures, they create a safe space for employees to explore new ideas and technologies. A culture of innovation can attract and retain top talent, people who are eager to contribute their creative insights and problem-solving skills. Failing fast also promotes a growth mindset within the organization, where failures are viewed as valuable learning experiences that contribute to future successes.

How many retailers do you know that celebrate learning from failures?

Customer Centricity

In our work with retail organizations, we have found the core of a retail innovation doctrine should rest on a deep understanding of and focus on the customer. This can involve conducting market research, gathering customer feedback, utilizing data analytics to gain insights into customer preferences, behaviors, and pain points, and developing an understanding of the shopper journey. By continually seeking to understand and anticipate customer needs, organizations can drive innovation that is aligned with customer expectations and deliver exceptional experiences.

---

* Bradley Pallister, "How Does Walmart Foster a Culture of Innovation?" Innovolo Group, March 13, 2023, https://innovolo-group.com/innovation-en/innovation-insights-en/how-does-walmart-foster-a-culture-of-innovation/.

Digital-native companies prioritize personalization and customer-centricity in their approach to customer engagement. They leverage data analytics and advanced technologies to understand customer preferences, behaviors, and purchase history. This enables them to deliver personalized experiences, recommend relevant products, and tailor marketing messages to individual customers. Digital-native companies focus on building long-term relationships and loyalty by continuously analyzing customer data to refine their offerings and provide superior customer experiences.

Traditional fast-moving consumer goods retailers often pay lip-service to customer centricity, stating that the customer is the most important part of the business all while directing resources to product procurement and merchandising in search of maximizing brand marketing funds. And while a growing number of retailers are employing marketing personalization, there is little to no added financial benefit to shoppers as retailers either "personalize" the hundreds of mass promotions available to any shopper or look to brand vendors to fund personalized deals.

Competitive Advantage

A retail innovation doctrine should highlight the importance of staying ahead of emerging technologies and trends. This involves actively monitoring technological advancements, conducting research and development, and piloting new technologies. Organizations should create a framework for evaluating and implementing relevant technologies that can enhance operational efficiency, improve the customer experience, and drive business growth. By changing paths, making the leap to ride technology's exponential growth curve, retailers can create the future of retail.

In today's retail landscape, technological innovation often serves as a key differentiator among competitors. By constantly testing and iterating on new technologies, retailers can be early adopters of cutting-edge solutions that meet customer needs and expectations. This gives them a first-mover advantage in the market, helping them attract customers, enhance brand perception, and drive business growth.

Military Doctrine / Tactical Principles: *These are specific guidelines for conducting operations at the tactical level, including the use of formations, maneuvers, and the employment of weapons and equipment.*

Retail Innovation Doctrine and Tactical Principles

Agility and Speed

Agility and speed are typically not words that come to mind when describing the fast-moving consumer goods retail industry. And yet that is just what's needed in a world of ever-accelerating tech-fueled innovation. As has been discussed earlier, traditional retail organizations are increasingly misaligned with the world past the inflection point. Retailers would be well served to transform and disrupt themselves before a faster moving competitor does it for them.

Look to transform existing decision-making processes, resource-allocation decisions, technology decisions, and more. Too often we see retailers eliminating layers of bureaucracy driven by a need to lower costs. Retailers would be well-served to eliminate structural, organizational, barriers to faster decision making. *Move faster!*

Failing Fast

Failure in nearly any initiative is seen as a waste of resources by traditional retailers. Given the thin profit margins that fast-moving consumer goods retail works on, it is no wonder that retailers are risk-averse. And this embedded risk avoidance is diametrically opposed to how digital-native companies think and how retailers must act going forward.

In the rapidly evolving landscape of retail technology, the ability to innovate and adapt quickly is crucial for staying competitive. One essential principle that drives successful innovation in this realm is the concept of failing fast. Failing fast means embracing the mindset that encourages rapid experimentation and learning from failures early in the innovation process. In successful innovation, it isn't if you're going to fail, it is when and how often. The occurrence is inherent with the commitment to rapid growth and innovation.

Kroger is one of the two traditional retailers (the other being Walmart) that I feel are furthest along in their evolution. Kroger has long focused on innovation—remember their involvement in the development of the barcode and scanning decades ago?—and overcoming the idea of failing fast has been key to their advancement.

Dan Whitacre is senior director of Kroger Labs and Technology. "A critical skill we've worked on with Deloitte is how to fail fast. They have their own innovation labs and understand how to take big challenges and break them into smaller ones. Teaching us to fail more quickly on the smaller problems enables us to avoid wasting time and instead focuses our energy on finding effective solutions to solve the bigger problems."*

Whitacre hits on a key point when he states that they've learned to "fail more quickly on the smaller problems and focus energy on the bigger problems." Retailers of any size would benefit from identifying smaller problems, identifying potential solutions, and working to implement and test those solutions quickly. Certainly, a company's size and resources impose constraints around so each must find their own way forward but the message here is to get moving.

Rapid Iteration and Learning

Innovation thrives when organizations embrace learning, experimentation, and rapid iteration. A retail innovation doctrine should promote a mindset that encourages exploring and testing new ideas, prototypes, and concepts. Organizations can create dedicated spaces or innovation labs to facilitate experimentation and provide resources for testing and piloting. Larger retailers can even consider using operating stores as "labs" where new innovation can be piloted in real world settings. By adopting an agile approach, organizations can quickly learn from failures, iterate on successful experiments, and continuously improve their products, services, and processes.

Proactive vs. Reactive Customer Engagement

Traditional organizations often have a transactional approach to customer engagement, viewing their business in terms of baskets and transactions. Digital-native companies prioritize customer-centricity

---

* Kroger, "Redefining the Grocery Customer Experience," March 21, 2022, https://technologymagazine.com/company-reports/kroger-redefining-grocery-customers-experience.

and aim for ongoing engagement; they intuitively think in terms of lifetime value. They are less concerned by individual sales and interactions and more concerned with building brand affinity and a long-lasting relationship with each customer, ensuring they'll return time after time due to brand loyalty, no matter what they're purchasing.

Digital-native companies take a proactive approach to customer engagement. They actively seek opportunities to initiate interactions with customers through personalized emails, targeted advertisements, and social media engagement. Digital-native companies use customer data and insights to identify customer needs, preferences, and pain points, allowing them to engage with customers in a more tailored and proactive manner. They also leverage chatbots and AI-powered technologies to provide instant and automated responses to customer inquiries, ensuring a high level of engagement and responsiveness. The ability to deliver seamless experiences across multiple touchpoints is critical and something many retailers don't do nor fully appreciate.

Traditional retailers often rely more on reactive engagement strategies. They respond to customer inquiries, complaints, or feedback primarily through traditional customer service channels such as in-store interactions or call centers, which tend to be less efficient and more cumbersome for customers to navigate or engage with.

Military Doctrine / Command and Control: *Doctrine outlines the command structure, decision-making processes, and communication systems that enable effective control and coordination of military forces during operations.*

Retail Innovation Doctrine and Command and Control

Organizational Structure

Traditional retail organizations tend to have hierarchical structures with multiple layers of management. Decision-making authority is concentrated at the top, and information flows vertically through the organizational hierarchy. Departments operate in silos, leading to slower communication and decision-making processes.

In contrast, digital-native organizations often adopt flat organizational structures. They aim for a more decentralized approach to decision-making, with fewer management layers and a focus on empowering employees. These organizations emphasize collaboration, cross-functional teams, and open communication channels. Flat structures allow for faster decision-making, greater agility, and encourage innovation and creativity across the organization.

Valve Corporation is a renowned video game and software development company known for its flat organizational structure. Valve is structured with a minimal hierarchy, and employees have a high level of autonomy and decision-making power.

Structure: Valve has eliminated formal titles and managerial positions. Employees choose which projects they work on and have the freedom to move between teams based on their interests and skills.

Decision-making: Teams are autonomous and make decisions collectively. There's an emphasis on collaboration and idea sharing.

Example: An employee can work on a gaming project one day and shift to virtual reality development the next based on personal interests and skills.

Now contrast Valve with a more traditional company, General

Electric. General Electric is a multinational conglomerate operating in various industries, including aviation, healthcare, and energy. It is an example of a company with a more traditional hierarchical organizational structure.

Structure: GE has a well-defined organizational hierarchy with clear reporting lines. It consists of multiple levels of management, from frontline employees to executives.

Decision-making: Decision-making typically follows a top-down approach, where major decisions are made by executives at the top and then cascaded down to lower levels.

Example: A product development decision may start with senior executives defining the strategy, followed by middle management setting specific goals, and then operational teams implementing these goals.

Examples of flat, nimbler, organizations abound. Amazon's "two-pizza rule" was made popular by Jeff Bezos, who limited the size of project teams to one that could be fed with two pizzas. The purpose is to encourage smaller, more agile, and efficient teams.

Military Doctrine / Intelligence and Reconnaissance: *Doctrine addresses the collection, analysis, and dissemination of information to support situational awareness and decision-making on the battlefield.*

Retail Innovation Doctrine and Intelligence and Reconnaissance

Awareness of New Butterflies

In a world of more new innovation "butterflies" every day—any one of which can disrupt your business—its mission critical to have awareness of new technologies coming into the market. There are many different ways to accomplish this; we'd suggest devoting time to the topic of new technologies in executive and management meetings that already take place.

Forward-looking associates at any level may be aware of new capabilities that should be on the company's radar. This could be a good opportunity to involve frontline workers, often younger and more tech-savvy, in efforts to understand what competitors are doing.

Competition

It's also highly recommended to keep close watch on both traditional retail competitors and new entrants into the retail space. Watch for new capabilities, new shopper engagement tools, etc.

There is no simple way to do this. I'd suggest reading both retail- and technology-related newsletters, talk with your trusted vendors for their insights, and leverage resources like FMI's Technology Directory. This could also be an opportunity to involve younger, more digital-native associates, asking them to periodically visit competitor's stores and websites to look for new capabilities.

Military Doctrine / Logistics: *Doctrine includes principles and procedures for providing logistics support to military forces, including supply, transportation, maintenance, and medical services.*

Retail Innovation Doctrine / Logistics

Innovation Process

In our work with retail companies like Hy-Vee, KVAT Food City, Wakefern, and others, we've developed a process for helping organizations discover relevant tech providers, doing due diligence on young solution providers, and engaging with company leadership. Having a process like this benefits everyone and helps embed the practice in how the organization operates.

Focusing on relevant innovation and having the leadership team together in one place focused on the same capability dramatically speeds up decision making around moving forward, or not.

Beyond simple awareness of new technologies coming into the market, retailers would be well served to create and follow an innovation process that includes staying aware of new tech, understanding of new capabilities, piloting new innovation that adheres to your strategy, and making quick decisions to terminate the pilot or move forward with deployment.

Cost and Resource Efficiency

Failing fast helps retailers avoid investing significant time and resources into ideas or technologies that may not deliver the desired outcomes. By identifying failures early, retailers can redirect their efforts and resources toward more promising opportunities. This approach saves valuable time, money, and effort that can be allocated to other innovative projects. The ability to course-correct swiftly based on early failures helps retailers optimize their resource allocation and maximize their return on investment.

Measure and Track Innovation Success

To ensure the effectiveness of innovation efforts, a retail innovation doctrine should emphasize the need to measure and track innovation success. Key performance indicators (KPIs) should be established to evaluate the impact of innovation initiatives on business objectives. Organizations can measure metrics such as revenue growth from new products or services, customer satisfaction scores, or employee engagement in innovation activities. Regular evaluation and reporting of these metrics will provide insights into the effectiveness of innovation strategies, enabling organizations to make data-driven decisions and course corrections as needed.

Military Doctrine / Training and Education: *Doctrine provides guidelines for training and developing military personnel to ensure they possess the necessary skills, knowledge, and mindset to carry out their assigned roles.*

Retail Innovation Doctrine / Training and Education

Talent and Skill Sets

The structural differences between traditional retail organizations and digital-native organizations reflect in the talent and skill sets required within each. Traditional retail organizations search for employees with expertise in store operations, merchandising, and supply chain management. These organizations often valued experience in the industry and domain-specific knowledge.

Digital-native organizations, on the other hand, prioritize skills in technology, digital marketing, data analytics, and user experience design. They seek talent proficient in emerging technologies, agile methodologies, and customer-centric approaches. These organizations often attract employees with a passion for technology, innovation, and a desire to disrupt traditional retail paradigms.

Military Doctrine / Joint and Combined Operations: *With the increasing importance of multinational and interagency cooperation, doctrine addresses the coordination and integration of military forces from different services or countries.*

Retail Innovation Doctrine / Joint and Combined Operations

Community Building

Traditional retailers have recognized the importance of community building and user-generated content in customer engagement. They may incorporate elements like loyalty programs, customer feedback initiatives, or user-generated content campaigns to foster customer participation and strengthen brand-consumer relationships. However, digital-native companies take these to a deeper level.

Digital-native companies often emphasize community building and user-generated content as part of their customer engagement strategies. They create online forums, social media groups, or interactive platforms where customers can connect, share experiences, and provide feedback. By fostering a sense of community, digital-native companies deepen customer engagement and build a loyal customer base. They actively encourage user-generated content such as product reviews, testimonials, or social media posts, which not only enhance engagement but also amplify brand advocacy.

Collaboration and External Partnerships

Collaboration is crucial in driving innovation. A retail innovation doctrine should emphasize the importance of collaboration within the organization and with external partners. Organizations can foster collaboration by breaking down silos and promoting cross-functional teams. Additionally, building strategic partnerships with startups, technology providers, and industry experts can bring fresh perspectives and drive innovation through knowledge sharing, co-creation, and access to external resources.

Wesley Rhodes is vice president of Technology Transformation and R&D for Kroger; Kroger Labs is the company's research and development division. "That collaboration ecosystem is some of the secret sauce impacting how Kroger Labs works," says Rhodes. "We look for partners that can seamlessly integrate in what we do, trusting them to bring the right kind of skills, to jump in and contribute to what we're trying to do. In this respect, Fusion Alliance here in Cincinnati has been an integral partner, complementing our staff, processes, and methods."* Kroger collaboration partners also include a number of universities and colleges where Kroger personnel work directly with faculty and students. As I've mentioned, Kroger goes so far as to have space on both the University of Cincinnati and Northern Kentucky University campuses not far from Kroger headquarters.

We have found developing ongoing relationships with retailers

---

* Ibid.

to be very valuable for both sides. While periodic (usually annual) innovation programs are the foundation, having an understanding of what the retailer is focused on, and where they see challenges and opportunities, enables us to regularly engage to share relevant, new capabilities that come to our attention.

Chapter Summary

- Retailers can use the wide-ranging elements of military doctrine as a framework for developing a doctrine appropriate to retail innovation.
- Retail innovation doctrines should include various elements of strategic concepts, tactical principles, command and control, intelligence and reconnaissance, logistics, training and education, and joint and combined operations.

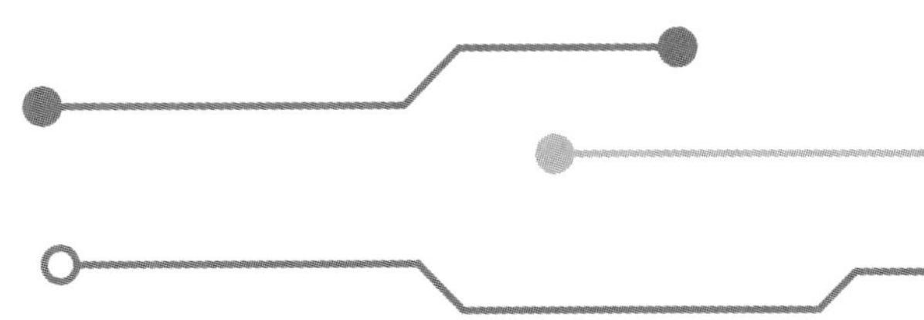

Part 3

CREATING THE FUTURE

"Make it so."

—Captain Jean-Luc Picard

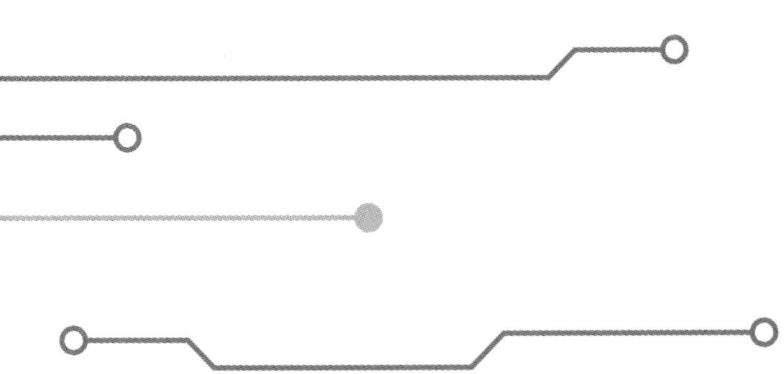

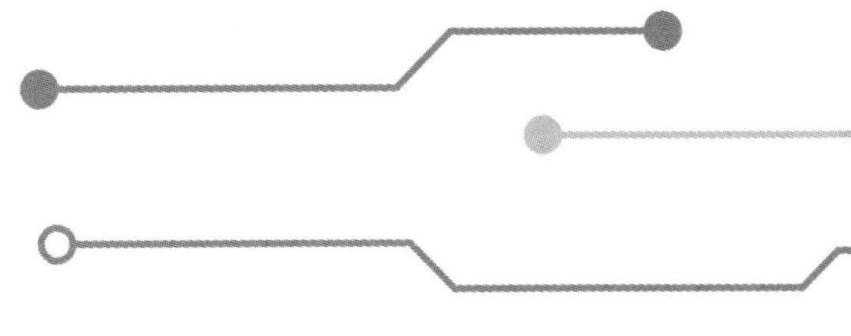

Chapter 10

WHEN ELEPHANTS FIGHT

When elephants fight, it is the grass that suffers.
—An ancient proverb of the Kikuyu people, a tribal group in Kenya, Africa, is as true today as when the words were first spoken, perhaps thousands of years ago.[*]

Creating the future of retail takes resources. Resources to learn about and monitor the growing number of new technologies—butterflies—coming into the market, any one of which could be disruptive to your business. More resources to pilot different capabilities in different areas across the enterprise. Resources to evaluate each pilot, making a decision to terminate or deploy. Even more resources to deploy chosen solutions, training associates to learn new capabilities, and more. And all of this at a faster and faster pace as the innovation flywheel accelerates up and out

---

[*] Eric Walters and Adrian Bradbury, *When Elephants Fight* (Victoria, BC: Orca Book Publishers, 2012).

the exponential growth curve.

The very largest companies have the resources to pursue multiple innovation initiatives in parallel across their organizations. But beyond resources, it takes a different organizational structure, new processes, a new culture, and the tech-first thinking that we've read about. There is a complex, integrated approach required to make this work.

In fast-moving consumer goods retail, Amazon, Walmart, and Kroger are in a class of their own when it comes to having resources to compete for the future. When looking at a list of the largest food retailers, the only others with annual revenue over $100 billion are Costco, Walgreens, Target, and CVS Health. And while those retailers have significant revenue, they do not appear to be as focused on leveraging technology innovation to create the future.

As you read through what Walmart, Kroger, and Amazon are doing, keep this thought front and center: These companies have the resources to build artificial intelligence (AI) systems across their organizations. AI, especially when it's tailored to a specific company's needs and goals and trained on the company's own data, is incredibly powerful. Look no further than Amazon and Walmart: According to Supermarket News, "artificial intelligence has the potential to bring in over $580 billion in extra profits for Amazon and Walmart according to new research from global research and advisory firm IHL Group."*

"Behind Walmart and Amazon for financial impact are CVS, Costco, Kroger, The Home Depot, and Target. When it comes to overall AI readiness, the same two top the chart, but the rest of the list varies slightly, with eBay and Rakuten joining the ranks for readiness (replacing CVS and Costco).

"AI is already transforming the retail market behind the scenes with traditional AI/ML improvements. Generative AI simply adds rocket fuel to that potential financial impact," said Greg Buzek, president of IHL Group."†

Watch for these companies to begin opening fast-growing performance gaps versus slower-moving competitors over the next couple of years. This kind of initiative is what creating the future is all about—and something most other retailers will have a difficult time keeping up with.

Amazon

In a very real sense, Amazon embodies the myriad battlefronts that define retail competition today. From automation to AI, from

---

* Alarice Rajagopal, "Amazon, Walmart Projected to Bring in Most Profits from AI," Supermarket News, August 18, 2023, https://www.supermarketnews.com/technology/amazon-walmart-projected-bring-most-profits-ai?NL=SN-02&Issue=SN-02_20230821_SN-02_652&sfvc4enews=42&cl=article_2&utm_rid=CPG06000000263787&_mc=em_SN_News_Supermarket News Daily_News_NL_08212023_50492&utm_campaign=56739&utm_medium=email&elq2=d78e83f78f0044e090a48c9176119dc8&sp_eh=e67ed679c944c6aeecc5d56dcdafbc8b77e72a779e514a0169520e2592b84d98.

† Ibid.

developing new revenue streams to sophisticated marketing personalization, and a disruptive new digital operating business model, Amazon is pushing the boundaries in multiple dimensions.

Over a decade ago, in 2012, Amazon acquired Kiva Systems, the robotics company that Amazon has relied on to help automate its massive distribution systems. And the company's automation efforts didn't stop there. As part of its Q1 2023 results, Amazon announced that it achieved an automation milestone: "1 billion packages handled by Robin, a robotic system used across Amazon's operations network in the U.S., Canada, and Europe. Robin robotic systems use artificial intelligence, computer vision, and machine learning to help employees handle and sort customer packages before shipment. For employees, this reduces the number of repetitive tasks, improving safety and helping them focus their time and energy on activities that make better use of their talents."*

Amazon is aggressively expanding into healthcare through its acquisition of Pill Pack, getting the company into the pharmacy business, and, more recently, its acquisition of One Medical, moving Amazon into providing actual healthcare services.

Amazon was the first retail company to leverage its massive online audience into developing its own retail media network (RMN), creating a new revenue source generating an estimated $38 billion in 2022. The company continues to grow its Subscribe & Save subscription program, automating the purchase of routine products, and effectively removing those products and categories from

---

* Amazon, "Amazon.com Announces First Quarter Results," April 27, 2023, https://ir.aboutamazon.com/news-release/news-release-details/2023/Amazon.com-Announces-First-Quarter-Results/.

competition by competitors. To gain a perspective on the breadth of Amazon's activity, one has to look no further than a news release announcing first quarter (2023) results:*

- Continued to delight customers with convenient delivery options and broad selection. Nearly 26 million customers ordered items with Same-Day Delivery in the quarter, an increase of 50 percent compared to last year.
- Increased selection in its U.S. online store with the addition of new marquee beauty, home, and fashion brands, including Lancôme, Shiseido, World of Martha by Martha Stewart, Rent the Runway, KNOW Beauty by Vanessa Hudgens, as well as new The Drop collections from Harlem Fashion Row and Coco & Breezy. The Drop offers customers access to limited-edition, made-on-demand clothing collections by fashion influencers.
- Launched a new Fulfillment by Amazon (FBA) capacity management system to help sellers who use Amazon's storage, packing, and shipping fulfillment service to manage their inventory and capacity more efficiently as they scale their businesses. The management system provides sellers with more insight, predictability, visibility, and control to better plan and manage their inventory and supply chains, as well as increase capacity limits when needed.
- Acquired One Medical, a primary care organization offering both 24/7 virtual care services and in-office visits across the U.S. for preventive and everyday health, chronic care

* Ibid.

management, pediatric care, and mental health services. For a limited time, One Medical is offering annual memberships to new customers at the discounted price of $144 for the first year (regularly $199/year), the equivalent of $12 per month.

- Announced new AWS tools that make building with generative artificial intelligence (AI) easy, practical, and cost-effective for organizations of all sizes.
- Received Zoox's driverless testing permit from the California Department of Motor Vehicles, marking a key step to commercializing the self-driving robotaxi for the general public. With this permit, Zoox completed an initial run of its shuttle service for employees at its headquarters in Foster City, California, making it the first purpose-built robotaxi to operate on public roads in California.
- Expanded the Ring lineup with new devices. Ring Battery Doorbell Plus offers improved HD video and energy-saving features, and Ring Peephole Cam is an apartment and renter-friendly video doorbell that easily installs over an existing door peephole.
- Expanded the Amazon-built TV lineup with the Fire TV 2-Series, which makes getting an Alexa-enabled TV more affordable than ever, and new sizes of the Omni QLED Series.
- Unveiled three customer terminal designs for Project Kuiper, Amazon's low Earth orbit satellite broadband initiative. Kuiper's mission is to bridge the digital divide by providing fast, affordable broadband to communities worldwide that are unserved or underserved by traditional technologies. To use Kuiper's broadband, residential and commercial customers

will install an outdoor antenna, called a customer terminal, to communicate with satellites overhead. Amazon is preparing to launch two prototype satellites to test the entire end-to-end communications network this year and plans to be in beta with commercial customers in 2024.

- Opened Amazon Sidewalk to developer testing for the first time as part of the company's vision to put a billion Internet of Things (IoT) devices on the network. Amazon Sidewalk is a low-bandwidth, long-range community network that connects devices across long distances where Wi-Fi and Bluetooth signals often cannot reach. With Sidewalk coverage extending to 90 percent of the U.S. population, developers can now build and connect a new range of devices for the network.
- Launched a loyalty linking capability for Amazon One, Amazon's palm recognition service that lets customers enter, identify, and pay. Panera is the first restaurant chain to offer the new loyalty linking capability, allowing customers to easily enroll in and redeem rewards, and pay for their purchases with just their palm.
- Ranked No. 2 on *Fortune* magazine's World's Most Admired Companies list for the seventh year in a row.

These initiatives are just a subset of Amazon's activity in 2023's first quarter. And lest anyone discount Amazon's ability to compete in grocery, and their impact on the industry, consider these five things published (August 2023) by Scott Moses, partner and Head of Grocery at Solomon Partners:*

1. Amazon is the number five U.S. grocer (and soon to be number four). There are 600+ Whole Foods, Amazon Fresh, and Amazon Go grocery stores. They employ an estimated 1.1 million non-union U.S. employees.
2. Amazon has $65 billion estimated U.S. grocery sales. Amazon has had more than 2,000 percent U.S. grocery sales growth in the last twenty years and is the leading online grocer.
3. Amazon has more than 200 million Amazon Prime members and $139 annual prime subscription fees generate $28 billion in cash each year. Amazon has more than $30 billion in annual advertising revenue.
4. Amazon has a AA credit rating and a nearly unlimited capacity for investment.
5. Amazon has a $1.4 trillion market value; this is more than ALL U.S. grocers *combined.*

* Scott Moses, "5 Things You Might Not Know," Solomon Partners, August 2023, https://solomonpartners.com/wp-content/uploads/2023/08/Scott-Moses-5-Things-You-Might-Not-Know-Amazon-Aug-2023.pdf.

Walmart

Walmart is the world's largest retailer, generating $572 billion in 2022 revenue and having an estimated 230 million shoppers visiting the company's stores and online properties each week. It was not until 2016, when Walmart acquired Jet.com, that the company embarked on transforming itself into a physical/digital hybrid retailer. What has been most impressive is how Walmart's leadership has transformed the company's culture, moving the importance of technology innovation to center stage for both its physical stores and online initiatives.

"In 2019, Walmart invested $11 billion in technology, an increase of 40 percent from 2018. This investment allowed the company to open new e-commerce fulfilment centers, develop new mobile applications for customers, and expand its online grocery delivery services. In 2020, Walmart reported over 1 million robots working across its stores and warehouses worldwide. Furthermore, the company has also implemented AI-powered chatbots on its website and mobile app to give customers faster responses to their queries."*

Walmart has continued to invest heavily in tech innovation, spending over $7 billion in 2022 on supply chain, omnichannel, and other tech. Following are just some of Walmart's initiatives . . .

Distribution Center Automation

Walmart invested in Symbotic, the brainchild of distribution savant Rick Cohen (owner of C&S Wholesale Grocery). Symbotic's

* Bradley Pallister, "How Does Walmart Foster a Culture of Innovation?" Innovolo Group, March 13, 2023, https://innovolo-group.com/innovation-en/innovation-insights-en/how-does-walmart-foster-a-culture-of-innovation/.

technology automates vast distribution centers and Walmart has publicly claimed the technology improves cost per unit by 20 percent.

Micro-fulfillment Centers

Walmart has an estimated 4,700 stores across the United States, located within just 10 miles of 90 percent of the U.S. population. Walmart acquired Alert Innovation in late 2022 and is in the process of deploying micro-fulfillment centers (MFCs) to a growing number of stores, turning the store into a local fulfillment center. The MFC's inventory is separate from the store's inventory to allow fine-tuning assortment and minimizing out-of-stocks for both in-store and online shoppers. Walmart is leveraging its massive stores into strategic assets as supply chains are transformed.

Drone Delivery

Building on testing drone deliveries for several years, Walmart has expanded its partnership with DroneUp, which will enable Walmart to reach 4 million households across six states. Walmart has stated that it could deliver more than 1 million packages annually by drone.* Further, Walmart is exploring offering drone delivery services to other local businesses, creating additional revenue streams as the company seeks to monetize the infrastructure it is building.

---

* Kim Souza, "The Supply Side: IGD Names Walmart's Top 5 Tech Innovations in 2022," Talk Business & Politics, January 21, 2023, https://talkbusiness.net/2023/01/the-supply-side-igd-names-walmarts-top-5-tech-innovations-in-2022/.

Automated Delivery Vehicles

Walmart is employing a growing number of autonomous vehicles used to deliver products between stores to shorten delivery times and reduce out-of-stocks.

InHome Delivery

As Walmart expands its supply chain capabilities through MFCs and drone delivery, it is expanding its InHome delivery service from 6 million households to a goal of 30 million households.* Through its InHome delivery service, Walmart will deliver your order to your kitchen, garage, or doorstep. This is in addition to shopper pickup at the store of online orders.

Membership Benefits

Walmart launched a rewards program with Ibotta in late 2022 that provides cash savings to shoppers. This builds on a partnership with Paramount to provide video streaming to Walmart members. Streaming benefits are building momentum through new partnerships with Spotify and the company's Netflix Hub at Walmart. Walmart is aggressively building out its Walmart+ Membership rewards program, increasingly positioning it as a viable alternative

---

* Sarah Perez, "Walmart to Expand InHome Grocery Delivery to 30 Million US Households in 2022," TechCrunch, January 4, 2022, https://techcrunch.com/2022/01/04/walmart-to-expand-inhome-grocery-delivery-to-30-million-u-s-households-in-2022/?guccounter=1&guce_referrer=aHR0cHM6Ly93d3cuZ29vZ2xlLmNvbS8&guce_referrer_sig=AQAAAKy4mSwFwe7nelMj9r72XxDs25VZdrOyk-Qc2QILGjhQ7oB5dZ8r2f-WCa-tdVWmCCaahM8reU2KFbAdZdt294B4DTTh_ZaXAKyZ14hFY6NUkt6QWUSZ1r7OPxJm7hA8dWKGbNRSHYaFDVhyUfEoMA4kLUMzt3tUANHyd2g_TGjA.

to Amazon's Prime program and Kroger's Boost program.

Scan and Go

The app-based scan-and-go capability is positioned as a benefit for Walmart+ members. Approximately 25 percent of shoppers at Sam's Club use the scan-and-go payment technology.

Social Commerce

Walmart has already hosted nearly two dozen livestream shopping events across platforms including Twitter, YouTube, and TikTok. It also has launched Walmart Creator* that enables influencers to monetize products from the retailer by creating video-based content. The company is connecting its social commerce activities to Walmart Connect, the company's retail media network (RMN), offering brands and advertisers additional channels to reach the shopper.

Digital Shopper Engagement

In the past year, Walmart acquired Zeekit, a virtual try-on technology and added it to Walmart's app. Along with that came Choose My Model, all designed to help apparel shoppers online. Then came Text to Shop that lets customers text product reminders to their accounts for next time they're shopping.

---

* Kim Souza, "The Supply Side: IGD Names Walmart's Top 5 Tech Innovations in 2022," Talk Business & Politics, January 21, 2023, https://talkbusiness.net/2023/01/the-supply-side-igd-names-walmarts-top-5-tech-innovations-in-2022/.

Health

"Leveraging its scale and role as a community center, the retailer's healthcare ambitions took on fuller proportions in 2022. Expansion of its network of Walmart Health locations adjacent to Walmart Supercenters is expected to see sixteen new locations open by the fall of 2023, the company said. *

Artificial Intelligence

"Over the past six years, Walmart has gone from a handful of in-house data scientists to hundreds," according to Srini Venkatesan EVP, U.S. Omni Tech at Walmart Global Tech. These data scientists serve on teams related to supply chain forecasting, optimization, and labor/demand planning; search and personalization; as well as emerging technologies. "We are really spending a lot on internal development because we feel this is our competitive secret sauce."† AI permeates everything Walmart is doing, from supply chain activity to search and personalization to new capabilities like the virtual try-one app capability mentioned earlier.

Retail Media Network

Walmart built Walmart Connect from scratch over the past several years, leveraging the media network into a significant new revenue

---

* PYMNTS, "Walmart Joins Best Buy in Offering OTC Hearing Aids," October 17, 2022, https://www.pymnts.com/news/retail/2022/walmart-joins-best-buy-offering-hearing-aids/.

† Sharon Goldman, "AI Is Embedded Everywhere at Walmart," VentureBeat, June 9, 2022, https://venturebeat.com/ai/ai-is-embedded-everywhere-at-walmart/.

stream for the retailer. The company brings together online purchase data—*real-time data* from its nearly 5,000 stores, Walmart.com, and the Walmart App—to provide advertisers the ability to identify and target very specific audiences. Sitting on top of all that data are increasingly sophisticated machine-learning algorithms designed to provide advertisers with effective impact. Walmart is expanding its network to include in-store radio and sampling programs, adding to its impressive digital reach. Walmart Connect generated $2.7 billion in revenue in 2022 and is expected to generate more than $3 billion in 2023.

Augmented Reality

Walmart's innovation lab, Store No 8, is developing augmented reality (AR) capabilities to enrich the shopping experience, overlaying the physical world with digital information. Alongside AR for shoppers, the Store No 8 team is creating AR-based capabilities for store associates with an eye to improving productivity and customer experience.

Review this list of innovation initiatives Walmart is focused on again. And keep in mind these are just some of the public initiatives; we can safely assume there are dozens more being quietly pursued. But what I want you to grasp is the breadth of what Walmart is doing here, they are transforming every part of their massive organization, from supply chain to logistics to distribution to store operations to digital shopper engagement. Along the way they are creating massive new sources of revenue like Walmart Connect, monetizing the delivery infrastructure being built, and expanding into healthcare.

Walmart is clearly creating their future. And there are few retailers that can keep pace.

Kroger

Kroger is one of the few competitors that may be able to keep pace with Walmart's initiatives. And I firmly believe it is this competition that is the rationale behind Kroger seeking to acquire Albertsons. That play is not about lowering the cost of paper towels by a few cents, rather, Kroger is looking to bulk up, aggregating the financial and technological resources needed to keep pace with Walmart and Amazon as the future of retail is created.

Kroger has a long history of innovation. As read about earlier, Kroger tested the first prototype scanner in a Cincinnati store in 1972. In the 1990s, Kroger was quietly testing early frequent shopper (loyalty) capabilities in different stores.

So, let's take a look at some of the things Kroger is working on. Everyone in the industry reads about different projects or initiatives at different times in the trade press, but looking in totality at what Kroger is doing clearly calls out how the company is rapidly transforming itself for the future.

Automated e-Commerce Fulfillment

As has been well publicized, Kroger has a partnership with Ocado, the UK-based company that builds automated distribution centers for fulfilling online orders. Kroger now has eight Ocado "sheds" in operation fulfilling orders across fourteen states. The company is building out a hub-and-spoke model in different markets, including Texas where the primary Ocado fulfillment center is in Dallas and spokes are located in Austin and San Antonio.

Stealth Market Expansion

What is not fully appreciated is how Kroger is using the Ocado capability to enter new markets without a physical store presence. This strategy is in place in Florida where Kroger is quietly growing market share. One can speculate that it was Kroger's data that led to that strategy, the company knowing that a fair number of shoppers moved to Florida or spend time there and so the Kroger name is well known.

Drone Delivery

Kroger is continuing a pilot with Drone Express in the Cincinnati market.

Automated Delivery Vehicles

Kroger has a partnership with Gatik, a provider of autonomous delivery vehicles that are used to deliver eCommerce orders from the distribution center to several stores in the Dallas/Fort Worth market for customer pickup.

Kroger continues to work with Nuro in other markets, using automated delivery vehicles to move orders from distribution centers to stores and direct to shoppers' homes.

Retail Media Network

Like Walmart, Kroger has built an impressive retail media network (RMN). Leveraging detailed purchase data on over 60 million households, along with myriad vehicles and channels to the shopper, Kroger's Precision Media provides a powerful value proposition to advertisers. Expanding its network, Kroger has partnered with Roku and with Disney to provide targeted ads via streaming services. More recently, Kroger announced it is deploying Cooler Screen's digital door screens to 500 stores, expanding in-store ad opportunities. What's truly impressive is Kroger's ability to provide full attribution to advertisers: measuring views, clicks, opens, and ultimately the in-store or online purchase. No wonder Kroger generated over $ 1.2 billion* in 2022.

Data Science / 84.51°

If Kroger has a not-so-secret weapon it's 84.51°, the company's data science subsidiary and successor to the joint venture with dunnhumby that Kroger launched in 2003 to help the company create value from the loyalty data it was collecting.

Launched under then-CEO David Dillon, Kroger's Customer-First initiative transformed the company. The massive stores of customer-identified transaction data were mined to inform category product assortments at the store level, inventory levels, and pricing. The data also powered up Kroger's personalized promotions, typically a monthly communication of strategic, precision-targeted, promotions designed to reward shoppers and grow baskets. Those personalized promotions were largely credited with the amazing fifty-two consecutive quarters of same-store sales growth Kroger achieved that came to an end in early 2017.

The 84.51° team now applies data science to a wide range of activities across the organization. One area of focus has been health and wellness, as they build on Kroger's OptUP app that provides a general nutrition score for each food product sold. The data science

---

* Stephen Babcock, "Kroger's 4 Pillars of e-Commerce Growth," The Current, March 2, 2023, https://thecurrent.media/kroger-ecommerce.

team has been working on scoring foods to specific health conditions. That initiative is just one part of a larger push into healthcare, including partnerships with health insurance companies.

Data science continues to power up marketing personalization, digital engagement, and marketing/merchandising decisions across the company.

Machine Learning

The 84.51° team is also heavily focused on machine learning. Kroger "is rapidly developing a 'machine-learning machine' that can build and deploy very large numbers of models with relatively little human intervention."*

Vertical Farming

Kroger works with 80 Acres Farms, a company building out leading-edge vertical farms that can be located near stores or distribution centers. 80 Acres supplies Kroger with fresh produce year-round that Kroger offers in hundreds of stores.

Ghost Kitchens

Ghost kitchens provide a way for grocers to compete with restaurants as takeout and delivery expanded during the pandemic. Kroger has partnerships with Saladworks and Kitchen United in a growing number of stores. In some locations all the prepared foods are sold

---

* Tom Davenport, "84.51° Builds a Machine Learning Machine for Kroger," Forbes, April 2, 2018, https://www.forbes.com/sites/tomdavenport/2018/04/02/84-51-builds-a-machine-learning-machine-for-kroger/?sh=77b96c4464e1.

under one brand but in others the food offerings are from different restaurant banners.

Artificial Intelligence

Kroger partnered with NVIDIA, whose specialized chips are used to power artificial intelligence applications, to open a lab in Cincinnati to explore the use of AI across the Kroger enterprise.

Healthcare

Kroger is expanding into healthcare, building on its 2,200 pharmacies and over 226 Little Clinics, in-store clinics offering diagnostic treatments, ongoing health management, wellness, and preventative care.

Store Operations

Kroger has long used a system it developed to monitor the inflow of shoppers to the store and then project how many checkout lanes will need to be open in the next fifteen minutes. Next time you're in a Kroger store, look for a large screen typically mounted on the ceiling in back of the checkout area. While not perfect, my experience is that Kroger's front-end service as measured by time in the checkout line is superior to competitors like Safeway.

And Kroger is extending its AI initiatives into operations. "Kroger has collaborated with Google Cloud and Deloitte on two purpose-built applications to enhance associate productivity. Google and its partners were "able to create something that leverages [AI] to continue to learn, so that we're always helping associates know what the next thing that they should be doing is," explained Jose Luis Gomes, managing director, retail and consumer at Mountain View,

California-based Google Cloud, "and that is contextual and alive, so that if something happens, the prioritization changes."*

Rush Delivery

Kroger partnered with Instacart to offer thirty-minute delivery in many markets across the United States. So, as you can see, Kroger is also transforming its enterprise across everything from supply chain and distribution to products, store operations, and marketing. What is most impressive about what both Kroger and Walmart are doing is so much of this transformation is occurring in parallel. This requires not only significant resources but also a very sophisticated process to manage all these initiatives and apply learnings across the company.

While companies like Walmart, Amazon, and Kroger have massive resources they are able to bring to bear on creating the future, they also have process. They have built processes across their organizations to understand shopper's needs and wants. They have built processes to discover and learn about new innovation entering the marketplace. They have processes to govern what technology they decide to build themselves or what to partner with others on. And they have processes for decision-making around technology pilots and deployments.

---

* Marian Zboraj, "These 2 Retailers Are Projected to See Biggest Financial Impact from Generative AI," August 17, 2023, https://progressivegrocer.com/these-2-retailers-are-projected-see-biggest-financial-impact-generative-ai?utm_source=swiftmail&utm_medium=email&utm_campaign=PG_NL_TechTrend&mkt_tok=ODI1LUxTUC01NDUAAAG-NvS-9n9JiMeaM0SHm_ADERe2yc4FKEUoeEoThnM0tu6AdoIWmzIYs-2KT3XaAiSb5XB1GMj-_hFsFcXfyELIRcqD632cRWeIiYBvDqj4WX.

In short, what these companies have is a framework or methodology. Just what we've been talking about throughout this book. And it's having a methodology to follow in developing inspired innovation and committing to action that retailers need to be successful as technology accelerates. And that's particularly true for regional and independent retailers.

Competing for the Future . . . or Not

According to the National Grocers Association (NGA), there are 21,574 independent supermarkets* across the U.S. market; NGA defines an independent retailer as those companies that are privately owned by families or employees. Those stores represent an estimated half of all supermarket stores in the United States. FMI (the Food Industry Association) reports there are 45,575 U.S. supermarkets.† Those independent stores generate an estimated 33 percent of industry sales, or a little over $250 billion annually, again according to the NGA.

Those thousands of independent stores are served by a handful of wholesalers who supply a large portion of the product the independents sell and provide the retailer different services. The wholesaler is essentially an aggregator, combining product purchases of, let's say Chobani Yogurt, from the hundreds or thousands of stores they service, to buy at scale from Chobani and then deliver the ordered cases to the individual store.

---

* National Grocers Association, accessed September 2023, https://www.nationalgrocers.org/.

† The Food Industry Association, "Food Industry Facts," accessed September 2023, https://www.fmi.org/our-research/supermarket-facts.

Most wholesalers markup the product they resell to their retailers, providing a gross margin to cover costs. And for many years, those same wholesalers have relied on "inside margin" to fund operations and provide their bottom-line profit. Wholesalers are very creative in generating that inside margin, such as negotiating deals on product purchases from vendors and then only passing through a portion of the deal to their retailers.

Wholesalers also aggregate their retailers into different ad groups, negotiating on their behalf with vendors and brand manufacturers, creating advertising and promotion plans. And, as with retailers everywhere, wholesalers aggressively seek brand manufacturer marketing funds, some of which are passed through to the independent retailers, and some of which are kept by the wholesaler, adding to their inside margin.

So, it might appear that there's no way that independent retailers can effectively compete with the largest retailers given differences in scale and wholesaler costs. That's what the Robinson-Patman act of 1936 was intended to address, outlawing price discrimination, but the federal law was passed nearly a century ago in simpler times. The industry is infinitely more complex today than it was then; no one could have foreseen the myriad ways large retailers leverage their scale into advantageous terms. Further, no one has seriously pursued applying Robinson-Patman for decades.

While Robinson-Patman could supposedly address price discrimination in the form of product cost between the largest and smallest retailers, it does not extend to how the large companies are leveraging their size and scale into new revenue streams. And those new revenue streams at the largest retailers put even more pressure on the independent sector. At the 2023 Groceryshop show, Donna Tweeten, Hy-Vee's president, stated from the stage that Hy-Vee is building revenue from healthcare and retail media, and that she is not sure how a retail business would exist looking forward without other significant revenue streams.

The largest retailers have created their own advertising networks to access retail media dollars from brand advertisers, and small retailers are left behind. Lacking in scale, smaller retailers are forced to participate in third-party ad networks like Citrus or Adsta. And while these ad networks do provide some incremental revenue to the retailer, it is a small fraction of that seen by the larger companies. A dollar of ad revenue to Kroger or Walmart is wholly retained by the retailer. A dollar of ad revenue generated from a third-party network is split, some good portion kept by the provider (like a Citrus or Adsta), some going to the retailer, and some portion going to the wholesaler. By the time that dollar is split up there's little left for the retailer. So, while yes, there is incremental revenue, it's pretty negligible.

We are already seeing the largest retailers invest in applying artificial intelligence (AI) that's been trained to their own data and operations, creating growing performance gains. As we've discussed, AI is rapidly being integrated into existing optimization capabilities like labor scheduling and demand forecasting. But using AI to power up business process automation, replacing work and process done by people today with software systems, creates even further advantage in the form of cost savings, greater speed, and more effective decisions.

But creating the future takes massive resources. Think about what a company like Walmart or Kroger are investing in automating distribution centers. Even a small micro-fulfillment center can cost millions of dollars. Using innovation to create the future of retail is not cheap.

Traditional Wholesalers

Independent retailers often look to their wholesalers for technology guidance. And that's where the troubles begin. Wholesalers are very good at their core business: moving cases of product from brand manufacturers to independent retailers. Wholesalers have also been pretty good at developing ad plans, merchandising plans, and aggregating other retailer purchases—ad printing, shopping carts, etc.—to help their retailers obtain lower costs. From my observations and conversations though, wholesalers are not as adept at strategy. Especially strategy involving technology.

Too often we'll see some wholesalers simply bless certain solution providers based on what "rebate" is paid back to the wholesaler when a retailer buys a certain solution. There is not often a lot of due diligence or strategy behind that kind of guidance.

The other key issue is that often we'll see wholesalers develop a relationship with a technology provider based upon a recommendation from one of their retail customers. But in a world of exponentially growing butterflies, retailers—certainly smaller retailers—simply do not have the wherewithal to be aware of new capabilities coming into the marketplace. Nor will they understand how those new capabilities may converge to create other capabilities. A scattershot approach to new innovation will not succeed in creating the future.

But there are deeper issues.

Wholesalers have historically relied on inside margin for their profitability and became very adept at negotiating deals with brand manufacturers based on cases purchased. But as brand manufacturers began to focus on improving the effectiveness of their massive marketing spend—many of the largest consumer packaged goods (CPG) brands spend around 20 percent of their top line revenue on marketing—things began to change.

Brands saw that wholesalers were negotiating deals based on the number of cases to be purchased in a certain period of time, and then saw wholesalers either not pass along the full deal cost or only enable their retailers to receive the deal cost for a limited time, taking the extra deal money to their inside margin. Brand marketing funds going to wholesalers' inside margin don't help brand sales, so change was inevitable.

The first thing brands did was to tie special costs to actual product movement at retail—scan-downs. So, while the retailer could save $0.25 a unit on committing to some case quantity purchase, the brand would offer a $0.35 unit savings (for example) paid on how many units the retailer sold. But to get that added discount, the retailer would have to provide a movement report to the brand or vendor proving the number of units sold.

Needless to say, that while effective, the logistics of collecting movement reports across hundreds or thousands of stores was not a small effort. One would think this would have been a great opportunity for wholesalers to step into, collecting transaction data from their retailers each week and providing the reports to the brand manufacturers. But there were two problems: The first was that many independent retailers did not want their wholesaler to know their actual sales; it was a not uncommon practice for a retailer to purchase product from two or even more wholesalers. And second, wholesalers did not work too hard to collect the data, and that opened the door to a bigger problem.

Third-party buying groups—like Northwest Grocers and Alliance Retail Group—got their start by seizing the opportunity to aggregate independent retailers' transaction data to provide the scan-down reports to the brands. They then quickly moved into a position of negotiating deals on behalf of their retailer members, taking over what had been the wholesaler's role.

As data and technology became increasingly important, these groups moved to bring new services and capabilities to their retailers—increasingly relegating the wholesalers to simply moving cases. And wholesalers are continuing to miss opportunities. I can remember several years ago having conversation with C-suite wholesaler execs around the opportunity—or the need—to aggregate the sales data from their retailers and create a common digital infrastructure to provide capabilities around retail media. Unsurprisingly, wholesalers missed another opportunity.

Instacart recently filed to go public, in good part based upon the retail media revenue being generated across all the company's retailers and customers. "Advertising and other revenue is a large chunk of the company's sales, making up 28 percent of the company's total revenue for the first half of this year. That category of revenue is growing, with the company noting that it rose 24 percent year over year during the first six months of 2023, to $406 million."*

---

* Sam Silverstein and Catherina Douglas Moran, "Instacart Files Plans to Go Public," Grocery Dive, August 25, 2023, https://www.grocerydive.com/news/instacart-files-plans-to-go-public/691940/.

Instacart is positioning itself as a tech provider to smaller retailers, along with helping small retailers participate in retail media dollars. Instacart CEO Fidji Simo added that while Instacart works with thousands of companies, it is positioned to help smaller retailers keep up in an increasingly tech-driven industry. "Retailers understand that we invest more in technology custom-built for online grocery than any single grocer could—and that by partnering with Instacart, they can have the same technology edge as tech giants and startups, while also being able to focus on everything else it takes to run a successful business."*

So, wholesalers are late to the party again, letting Instacart develop the digital infrastructure stretching across thousands of independent retailers, and then leveraging that into growing retail media dollars, some of which are shared with the retailer. And, Instacart is growing that digital infrastructure, by providing retailers with smart carts and other technological tools to help them compete with the industry giants.

One more example: Instacart is offering a growing array of marketing tools to smaller retailers, including targeted marketing and personalization capabilities. Marketing personalization is quickly becoming a shopper expectation and the leading retailers have graduated to strategic hyper-personalization, growing their businesses one shopper at a time. Instacart is stepping up to help small retailers.

As the traditional mass-promotion-filled printed weekly ad goes

---

* Timothy Inklebarger, "Instacart Files Its Long-awaited IP Plans, with PepsiCo as Big Backer," Winsight Grocery Business, August 25, 2023, https://www.winsightgrocerybusiness.com/technology/instacart-files-its-long-awaited-ipo-plans-pepsico-big-backer.

away, increasingly replaced by personalized promotions relevant to each individual shopper, the ability to help smaller retailers is critical. Instacart, and even the groups like Northwest Grocers, are in a position to help their retailers gain access to CPG brand marketing funds as mass promotion evolves to personalized promotion. As this continues, wholesalers will be at risk of losing the ability to aggregate their retailers for merchandising programs, adding more pressure to margins. And wholesalers have growing risk as the future unfolds as we'll read about in upcoming chapters. A future reliant on digital platforms.

So, while the deck has long been stacked against independent retailers, the exponential world of retail represents a threat unlike any before. Smaller retailers, even multi-billion-dollar regional retailers, simply do not have the financial resources, associate skill sets, or the bandwidth necessary to create the future of retail. And wholesalers have a questionable record of helping their retailers navigate the future.

What to Do?

So, is the battle for the future lost before it's even begun for independent retail? Sadly, I do not have a crystal ball, but from where I sit, it's going to get ugly. So, is there anything to be done? There are a few things I'd suggest:

The first thing to be done focuses on education. Independent retailers must understand that they are now operating in a new environment, a world that has never before existed. A world of exponential growth in technology and a visceral understanding of what that means: constantly accelerating change. So many retailers—especially smaller retailers—simply do not understand the magnitude of the threat.

With understanding, independent retailers have a choice to make, a choice outlined in Chapter 5: Changing Paths. They either must make the leap to the exponential path or continue on the linear path into the future, knowing that the world will be accelerating around them.

And it's not just retailers that have to make this choice; the same choice confronts wholesalers. And from discussions I've had with C-level executives, there's no sense of urgency. For those executives, I'd suggest going back to the Innovation Flywheel and the discussion around the two things that make exponential growth dangerous: Early on its deceptive until it hits the inflection point and then just explodes. We're now just past the inflection point so buckle your seat belts.

Form a Skunk Works Project to Create the Future

For those retailers wanting to make the leap, there's work to do.

A "skunk works project" is loosely defined as an innovation project, involving a smaller team of people, that works outside the formal organization. The term originated at Lockheed Martin during World War II when the U.S. needed to rapidly develop a jet fighter to counter Germany's airpower.

What I'm suggesting here is that a small, core group of key retail executives partner with a small number of wholesaler execs in their own skunk works project to envision the future of retail. The work to be done follows the methodology I've mentioned earlier:

1. Work to identify restrictive practices and processes—and there are many of them in the retailer–wholesaler world that impede unified innovation. Identify them and then set them aside.

2. Create a vision for what you believe is the future of independent retail. And it's important to have set those constraining practices identified earlier aside to do this. Start with a clean whiteboard and brainstorm around if anything were possible, what would independent retail look like. Use the first-principles approach to work through creating that future vision.
3. Once that vision is created and agreed to by the skunk works team, commit to action. What is it going to take to make that vision a reality? What pieces need to come together? The team—retailers and wholesaler—needs to commit to making it happen.
4. Develop an innovation doctrine, initially for the project, and eventually to expand out to other retailers supplied by the wholesaler. And think of ways to embed that doctrine in how all of you operate.

One of the most important parts of this process is identifying, and then setting aside, processes, practices, and beliefs that constrain new ways of doing things. Because in the world of independent retailer, there's a lot of them.

Get on the Retail Media Bus

While I believe the retail media network bus has already left the station, there are still dollars out there to get. But the only way independent retail is going to realize any meaningful level of revenue is for the wholesaler to create their own media network extending across participating stores. And this means putting the systems, processes, and resources in place to build and operate that media network, just like the big retailers. As called out earlier, partnering with a third-party network solution simply divides up the available funds. Sure, less effort, but a commensurate loss in potential revenue.

To do this effectively means that, ideally, there is some uniformity to the digital infrastructure. Today, look across the independent retailers at any of the larger wholesalers and you'll see a plethora of e-Commerce platforms, email systems, website providers, apps, and more. That kind of splintered environment is anathema to building the powerful digital engagement platform needed to secure meaningful retail media dollars.

So, ideally all the participating retailers deploy the same digital e-Commerce platform so there is a common infrastructure for executing marketing campaigns. Importantly, I'm not saying that the retailers lose their identity, far from it. While they all have the same underlying platform, the retailers maintain their own branding across their websites, email, apps, and online shopping sites.

And that digital platform must provide a cohesive, seamless user experience across devices and channels. That's something that smaller retailers are still not fully appreciating. Amazon, Walmart, Kroger, and others all provide that seamless user experience. Far too often, mid-size and smaller retailers have disparate solutions they've tried to integrate and knit together—and that just does not work. This is something that many of the solution providers also don't understand—e-Commerce platform providers supplying online shopping but not providing the primary website, etc.

As an example, Stor.ai (stor.ai), recently acquired by and merged with Relationshop, is doing just what I'm suggesting. They are able to provide comprehensive digital engagement that has all been tightly integrated—website, email, app, and online shopping are all brought together in one platform.

Lowes Foods is another example. "Much of Lowes Foods' new omnichannel retailing approach is embodied in its recently refreshed app. While the previous version was built to power e-commerce, the new app is built for use across shopping occasions. The enhanced technology includes three modes that customers can toggle between—in store, pickup, and delivery—reflecting the different retailing channels shoppers use depending on their needs. Under the in-store tab, users can clip digital coupons and locate the exact store aisle where items are located. A barcode scanner pulls up product information. A list-building tool lets customers add and check off items with one click, and the tool can be shared across users."*

So, a common underlying digital platform enables the wholesaler to operate a retail media network (RMN) effectively and efficiently, selling ads and other digital promotions to brands that extend across all the participating retailers. But the effort doesn't stop there.

Data and Programmatic Ads

The largest retailers are leveraging massive amounts of data combined with programmatic ad serving to reach both shoppers and non-shoppers across their market areas with brand-funded

---

* Jeff Wells, "How Lowes Foods Uses Digital to Supercharge its Unique Shopping Experience," Grocery Dive, July 11, 2023, https://www.grocerydive.com/news/how-lowe-foods-uses-digital-to-supercharge-store-experience/685586/.

advertisements, with the promoted products only available at the retailer.

It's possible to secure information through data broker services on nearly every household in a given market area. This data could include demographic data—number and ages of people in the household, income, education, etc. It can also include other information like if there are pets in the household. Using that data, it's possible to target ads through programmatic ad servers that are attuned to that specific household. For example, the wholesaler/independent media network could partner with Purina to advertise a new cat food to those households with cats, with the ads popping up when the consumer is online.

And there's more. Refinition (refinition.com) can help retailers understand where else their customers are shopping, able to anonymously track most of the smartphones that shoppers bring into a given store and then monitor what other retailers they go to. That's powerful information that can be incorporated into marketing campaigns to further fine-tune efforts.

Personalization and Changing Business Models

In early 2023, Strack & Van Til, a twenty-one-store retailer outside Chicago, joined Coborn's, Weis Markets, Harp's, and other regional retailers in implementing Birdzi's automated, strategic, hyper-personalization capability to deliver personalized savings to their shoppers. These innovative retailers are changing the traditional model of always relying on consumer packaged goods (CPG) brands for funding discounts. Instead, Strack & Van Til and the others are effectively leveraging their entire store product catalogs as offer pools, and funding personalized discounts, to maximize relevancy and meaningful financial value to each of their shoppers.

This expansion to using all the SKUs in the store and the retailer funding the discounts represents the next wave of personalization in retail, and an area that wholesalers are well positioned to assist and enable their retailers in deploying.

The latest personalization capabilities, like Birdzi, leverage artificial intelligence (AI) to create a custom strategy to grow the value of each shopper, and then create the best mix of offers and savings to achieve that strategy. Beyond growing basket size and visits, this approach is also highly successful at growing shopper engagement across the store, increasing the number of categories shopped regularly.

Wholesalers structured as co-ops, companies like Associated Wholesale Grocers, are well positioned to assist and enable their retailers in pursuing these personalization strategies, and helping their smaller retailers access sophisticated capabilities.

The Store as a Lab

The wholesaler/retailer partnership can provide some unique advantages. Imagine different retailers testing and piloting different innovative capabilities in parallel to mimic the scale of the largest retailers. Properly organized, independent retailers could have the ability to simultaneously test various new innovations across dozens of stores, some focused on supply chain capabilities, others on store operations, and yet others on digital services and shopper engagement.

While an opportunity, pursuing the use of stores as innovation labs will take some doing. The wholesaler is in position to coordinate the activity across retailers, assist in discovering and evaluating new technologies to test, and compiling the results and learning to share across their networks.

Other Retailers

When looking at the scale of FMCG retail, there are myriad well-performing, successful regional retailers occupying the space between the largest retailers like Walmart, Kroger, and Amazon, and the wholesaler supplied independent sector. For purposes of discussion, let's think of this segment as "Tier 2 retailers."

I think it is this middle ground where we are going to see some of the most intensive battles over the future of retail being waged over the next several years. These multi-billion-dollar retailers have resources to deploy against new innovation, but they must be judicious in their pursuits.

Schnuck Markets out of St. Louis is a good example of a regional retailer methodically pursuing innovation. "Instead of embracing the latest fads in technology, Schnucks has systematically integrated smart tech into its operations, leveraging everything from electronic shelf labels (ESLs) to inventory monitoring robots."* The retailer has partnered with Instacart on Caper smart carts, pricing and

---

* Chris Walton, "Schnucks: Out-Innovating Amazon, Walmart and Everybody Else in Grocery," Forbes, July 30, 2023, https://www.forbes.com/sites/christopherwalton/2023/07/30/schnucks-out-innovating-amazon-walmart-and-everybody-else-in-grocery/?sh=406e37f224f4.

promotions capability, and more.

"Regional grocers are caught in a challenging middle ground. They often position themselves as hometown grocers, yet at the same time are expected to compete with the industry's biggest players. Regional chains operate stores in prime real estate that puts them in direct competition with Walmart, Kroger, and Albertsons, and they carry many of the same products. Regional retailers also have the money to invest in online grocery shopping, to develop new apps, and to scale technologies that help them operate more efficiently.

"But regional operators don't have the same capital advantages and capabilities as the big names. As technology evolves and as consumers' digital habits change, the gulf between regional grocers and national players threatens to grow. The proposed Kroger-Albertsons merger, which is currently working its way through the regulatory process, would only add more pressure."*

Hy-Vee, a top ten U.S. supermarket retailer, headquartered in Des Moines, Iowa, is working hard to keep pace. In the first half of 2021, the retailer launched Hy-Vee Red Media, its retail media network. Leveraging in-house data and capabilities, along with strategic partnerships, Red Media enables CPG brands to reach shoppers on Hy-Vee's digital properties in addition to off property, using programmatic ad serving and other technologies.

Hy-Vee has also been a leader as it pushes into healthcare, building on the success of its in-store pharmacies. Hy-Vee operates a

---

* Jeff Wells, "Pardon the Disruption: Regional Grocers Face a Fraught Path to Digital Growth," Grocery Dive, July 10, 2023, https://www.grocerydive.com/news/pardon-the-disruption-regional-grocers-face-a-fraught-path-to-digital-grow/685690/.

specialty pharmacy and has recently opened infusion centers along with its own pharmacy benefit management company to provide service to other businesses.

One way to maximize resources might be for these retailers to leverage the share groups they participate in. Many of these groups are comprised of non-competing retailers and, like the wholesalers, may be able to coordinate testing of different technologies across member retailers.

While something like this may not have been appealing in the past, given the acceleration of technology and innovation—remember those butterflies?—these retailers are going to run out of bandwidth to keep up with the largest retailers and leveraging their share groups may be a way forward.

Automation as a Service

In what could be a game-changer for mid-market retailers, Rick Cohen's Symbotic is working to bring sophisticated distribution center automation to smaller players. Greenbox Systems, a new joint venture between Symbotic and SoftBank, will provide Symbotic's warehouse-as-a-service in both single and multi-tenant facilities. Based on Symbotic's performance in existing deployments, "GreenBox customers will be able to reduce inventory and costs while boosting SKU count and agility, all without associated capital expenditures and operational complexity."[*]

"Symbotic's end-to-end automation consists of a fleet of fully autonomous robots with advanced vision and sensing capabilities. These robots induct, store, retrieve, and palletize products in their native packaging at industry-leading throughput rates and with 99.9999 percent accuracy. Symbotic's AI-powered software orchestrates the work of hundreds of mobile robots to fulfill customer orders from within the high-density buffering and storage structure that flexibly handles virtually unlimited SKUs. The AI-enabled system builds mixed-SKU pallets with unmatched speed and efficiency, ultimately changing the way consumer goods are moved through the supply chain. The disruptive technology enhances warehouse density and capacity; increases inventory turns, reduces errors and waste; and enables faster fulfillment of customer orders. Currently, Symbotic systems service more than 2,600 stores for some of the world's largest retailers, grocers, and wholesalers, including Walmart, Albertsons and C&S Wholesale Grocers."[†]

The Symbotic announcement is a good example of how

---

[*] Business Wire, "SoftBank Group and Symbotic Establish New Warehouse-as-a-Service Joint Venture To Transform Supply Chain Services Through Automation and A.I.," Yahoo! Finance, July 24, 2023, https://au.finance.yahoo.com/news/softbank-group-symbotic-establish-warehouse-110000986.html?guccounter=1.

[†] Symbotic, "SoftBank Group and Symbotic Establish New Warehouse-as-a-Service Joint Venture To Transform Supply Chain Services Through Automation and A.I.," July 24, 2023, https://www.symbotic.com/about/news-events/news/softbank-group-and-symbotic-establish-new-warehouse-as-a-service-joint-venture-to-transform-supply-chain-services-through-automation-and-a-i/.

technology is moving ever faster. Prior to the GreenBox initiative, sophisticated distribution center automation was only available to larger retailers having the resources to deploy it. "GreenBox is the realization of a vision I've had for many years to bring AI-enabled automation to companies of all sizes," Cohen shared.*

Chapter Summary

- The exponential world brings with it challenges for small and mid-size traditional retailers who have limited resources. The key will be leveraging their ability to move and make decisions quickly along with bringing a laser-like customer focus to new innovation.
- Consider initiating a skunk works project to partner retail leaders with wholesale leaders to ideate on the future of retail. Both sides provide unique perspectives to pain points and possibilities, and the likelihood of developing mutually beneficial solutions is high.
- Leverage AI powered business process automation to multiply your resources.

---

* Amy Feldman, "Billionaire Behind Walmart's Warehouse Robots Gains More Than $7 Billion in a Day," Forbes, July 31, 2023, https://www.forbes.com/sites/amyfeldman/2023/07/31/billionaire-behind-walmarts-warehouse-robots-gains-more-than-7-billion-in-a-day/?sh=d39dbac231f2.

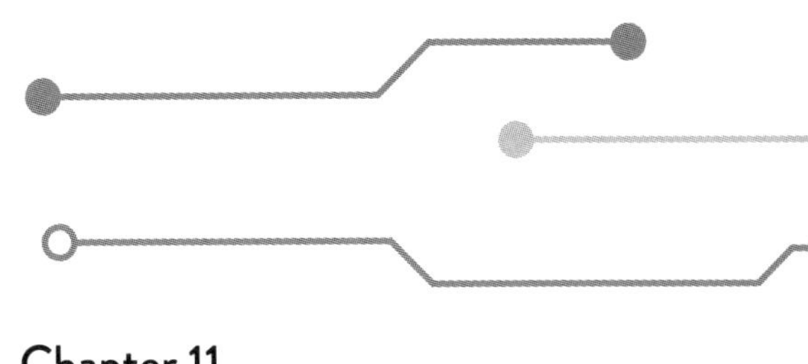

Chapter 11

RETAIL AS A PLATFORM

Think about retail today. Consider how a retail company is traditionally organized. There's the merchandising department where the category managers sit, making decisions around what new products to slot, which poor-performing products to delist, pricing, promotions, ad plans, and more. There's the marketing department, charged with producing and distributing the weekly ad, overseeing the company's loyalty program, new store opening campaigns, and so on. We have store operations, responsible for scheduling, cleaning, stocking shelves, rotating product, checking out shoppers; everything to do with running a store. There's the HR department, coordinating hiring, terminations, personnel policies, etc. Then we have the IT department, the real estate department, legal, and on and on.

Conway's law states that an organization's systems mirror the way the organization operates. That's certainly true in retail. Consider how a retailer is organized as described here. Each of those departments or operational areas inevitably have their own systems and data to support how they do things. The category managers have a system—or multiple systems—that provide them the data to make decisions around what products to carry, how to price them, and when to promote them. Marketing has their CRM system, and other tools to support what they do. Store operations have their task management systems, store labor scheduling systems, ordering systems, and more.

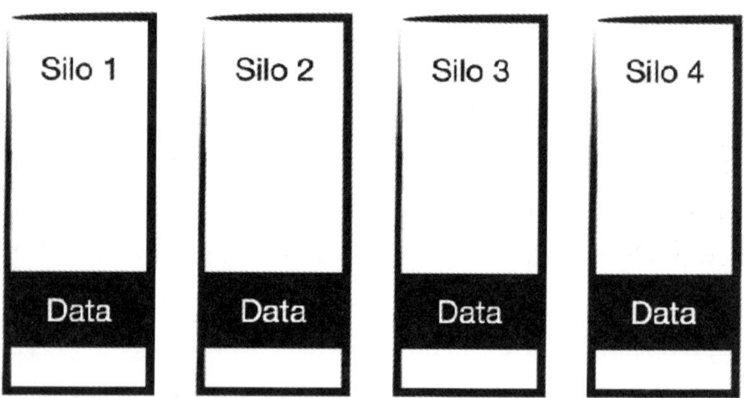

Again, mirroring the real world, those systems are typically separated, just like the merchandising team is located in their own office space, and the marketing group in their space. And just like the category buyer will wander down the hall to the marketing department to discuss an ad block, IT has built point-to-point integrations between different data sources and applications to serve specific needs.

The result of all that is growing complexity created by data and application silos, task-specific integrations, and inefficiency. To be blunt, many retailers' systems and processes have developed into a giant hairball, a byzantine collection of data, tech systems, and processes that becomes more inefficient as the company grows. This is certainly not a structure—physical or digital—that is conducive to the agility required in the exponential world.

Traditional retail systems 'hairball'

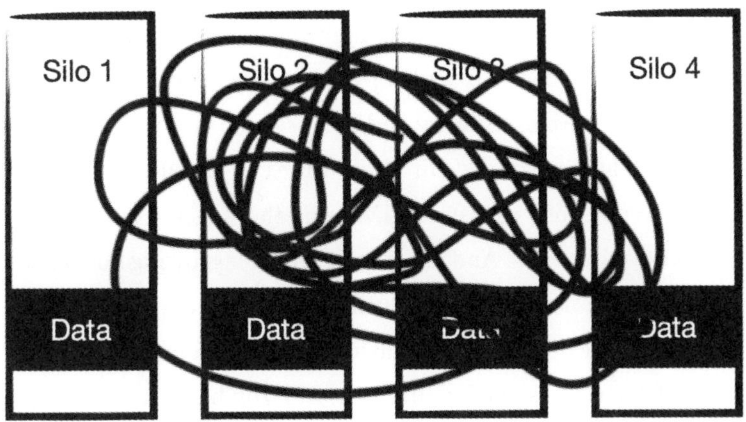

A New Model

The technology company Ant Financial Services Group was founded in 2014, and in 2019, just five short years later, served over 1 billion consumers. The company was spun off from Alibaba several years ago and uses artificial intelligence (AI) and data from Alipay, a mobile payments platform, to provide a wide variety of financial products. "The company serves more than ten times as many customers as the largest U.S. banks—with less than one-tenth the number of employees. At its last round of funding, in 2018, it had a valuation of $150 billion—almost half that of JPMorgan Chase, the world's most valuable financial-services company.

"Unlike traditional banks, investment institutions, and insurance companies, Ant Financial is built on a digital core. There are no workers in its 'critical path' of operating activities. AI runs the show. There is no manager approving loans, no employee providing financial advice, no representative authorizing consumer medical expenses. And without the operating constraints that limit traditional firms, Ant Financial can compete in unprecedented ways and achieve unbridled growth and impact across a variety of industries."*

We look at Amazon today and don't stop to consider it was not always a colossus. Founded in 1994, by the early 2000s, Amazon had grown significantly as it expanded into other businesses and categories beyond books. And, like nearly all other growing businesses, Amazon grew into a series of business lines and departments that became silos, each operating individually and integrating with other parts of Amazon as needed. In short, Amazon had become a large, unwieldy, traditional business. And it was coming apart at the seams.

In 2002, Jeff Bezos wrote what has become a famous memo in which he mandated a new systems approach:

1. All teams will henceforth expose their data and functionality through service interfaces.
2. Teams must communicate with each other through these interfaces.
3. There will be no other form of interprocess communication allowed: no direct linking, no direct reads of another

* Marco Iansiti and Karim Lakhani, "Competing in the Age of AI," *Harvard Business Review*, January 2020, https://hbr.org/2020/01/competing-in-the-age-of-ai.

team's data store, no shared-memory model, no back-doors whatsoever. The only communication allowed is via service interface calls over the network.

4. It doesn't matter what technology they use. HTTP, Corba, Pubsub, custom protocols—doesn't matter.
5. All service interfaces, without exception, must be designed from the ground up to be externalizable. That is to say, the team must plan and design to be able to expose the interface to developers in the outside world. No exceptions.
6. Anyone who doesn't do this will be fired.
7. Thank you; have a nice day!*

That memo triggered the transformation of Amazon from an online bookseller into a global digital platform. It didn't happen overnight. As a matter of fact, it took three major efforts until they got it right. But when they got it right, everything changed. That memo laid the foundation for Amazon Web Services (AWS), opening up to the outside world the technologies that power up Amazon internally, and created the digital foundation for the development of AI tools that power Amazon today.

The platform created by Amazon supports a rapidly growing number of business units, able to nimbly seize market opportunities. Beneath the surface, those myriad business units are nourished through a root system of shared data and information that continuously flows throughout Amazon. And that platform is powering enormous growth.

---

* Nordic APIs, "The Bezos API Mandate: Amazon's Manifesto for Externalization," January 19, 2021, https://nordicapis.com/the-bezos-api-mandate-amazons-manifesto-for-externalization/.

"Amazon (AMZN) is set to hit a major milestone in 2024: Becoming the largest retailer in the United States, according to JPMorgan analysts Doug Anmuth and Bryan M. Smilek. If this comes to pass, Amazon will be unseating Walmart (WMT) as the country's largest retailer. It would be a seismic shift, one driven by increased e-commerce penetration, faster delivery times, and the stickiness of Amazon Prime. JPMorgan estimates show that, in 2023, Amazon's gross merchandise volume, or GMV, will grow 11.6 percent year-over-over to $477 billion."*

Amazon's platform continues to grow in nearly every dimension. The company recently announced plans to use small businesses to deliver packages beginning later this year. Amazon Hub Delivery is intended to expand the company's last mile network through leveraging small local merchants like florists, bodegas, and others to do the final delivery of packages to shoppers.

> *"Machine learning is a form of artificial intelligence (AI) which mines data for patterns that can be used to make predictions. It took root at Amazon in 1999."*
>
> —*The Economist*

Meanwhile, in Amazon's distribution centers, next-generation robots are rolling out, including one called Sparrow that can pick individual products from containers and place them in bins—with

* Alexandra Garfinkle, "Amazon May Be the Largest U.S. Retailer in 2024, According to JPMorgan Analysts," Yahoo! Finance, June 25, 2023, https://finance.yahoo.com/news/amazon-may-be-the-largest-us-retailer-in-2024-according-to-jpmorgan-analysts-140145529.html.

near human-level dexterity. Think of the automation in Amazon's warehouses as an extension of its core digital operating platform. It's all about data and AI-powered automation.*

The Retail Digital Doppelgänger

Take a moment to picture your store. Delivery trucks pulling up to the delivery dock, product flowing into the backrooms and coolers. Store associates and direct store delivery (DSD) vendors bringing products out onto the sales floor to stock shelves and build displays. Shoppers flowing in the entrance, moving into the departments and aisles, pausing in front of various categories pondering a purchase. Shoppers with carts and baskets of chosen products moving to the front end for checkout and exiting the store. It is a nonstop, complex dance of products and shoppers.

Now, imagine what's flowing through the physical store is data, information about every product and every shopper, everything digitized. You're able to "see" the case of Gulden's Spicy Brown Mustard coming off the delivery truck and moving to aisle 3 to be restocked. "See" the local dairy vendor bringing in cases of cottage cheese that's on sale that week. "Seeing" each shopper as they enter the store, choose a shopping cart, reference their shopping list on their smartphone, and begin navigating the aisles. "Seeing" the products in the shopping cart that's checking out in lane 5. Able to "see" each product's inventory in real time, increasing as products flow

* Will Knight, "Amazon's New Robots Are Rolling Out an Automation Revolution," Wired, June 26, 2023, https://www.wired.com/story/amazons-new-robots-automation-revolution/.

in the backdoor and decreasing as products are checked out and leave the store.

And there's more. You're also able to "see" the temperature of every refrigerated and frozen cooler and display case in the building. "See" the flow of shoppers up and down every aisle and through every department, along with which shoppers pause where to pick up a product and inspect it closely. Able to "see" how much energy the lights are using, the refrigeration system is consuming, and other equipment around the store, all in real time.

Now let's look at what's happening at the head office. The category managers going about their business making decisions around what products to delist, new products to slot, what to promote when. And taking the occasional coffee breaks, trips to the restroom, spending a few minutes talking with a colleague about the weekend, and then heading home at the end of the workday, all the lights turned off until the next morning.

The same thing happening down the hall in the marketing department, over in accounting, and upstairs in HR. A hive of activity . . . sometimes.

But now let's shift to the digital view of the world. Instead of all those people there are virtual AI agents "sitting" in their places. And those AI agents working around the clock, never stopping. Making better, more effective, decisions at light speed, drawing on more data and information than any human can even contemplate. And beyond the AI agents focused on specific tasks in each operational area, there's another level of AI agents looking across the entire enterprise, tweaking and optimizing every operation, constantly learning from the data.

Retail as a Platform

A platform can be defined as an application technology system that serves as a base from which a service is provided. Importantly, a platform is not just a tech system. A platform, as we're discussing it here, is a business model that reduces transaction costs and can support and enable innovation. A platform creates value by facilitating exchanges between groups, like product vendors and shoppers. Importantly, a *digital* operating platform can scale in ways traditional physical businesses cannot.

But creating retail as a digital operating platform on top of the historical, splintered, data architecture that exists in much of traditional retail today is a disaster in the making. Retail needs a new model. As we move up and out technology's exponential growth curve, the traditional retail organization and operating model with its specialized and siloed processes and data is increasingly unfit to compete with digitized retail platforms powered at their core by AI.

Now, here is where a lot of readers are going to sit back and say that it's not possible. Fast-moving consumer goods is too dynamic, too complex, and needs humans making key decisions—just like it's been done for the past 100 years. Until it's not.

Digital operating platforms already exist at Amazon, at Ant Financial, and other firms that have transformed themselves into AI-powered digital platforms. And those companies are changing the world, enabling new services, creating new shopper expectations, and powering massive value creation.

Retail as a platform is here . . . and it's evolving fast.

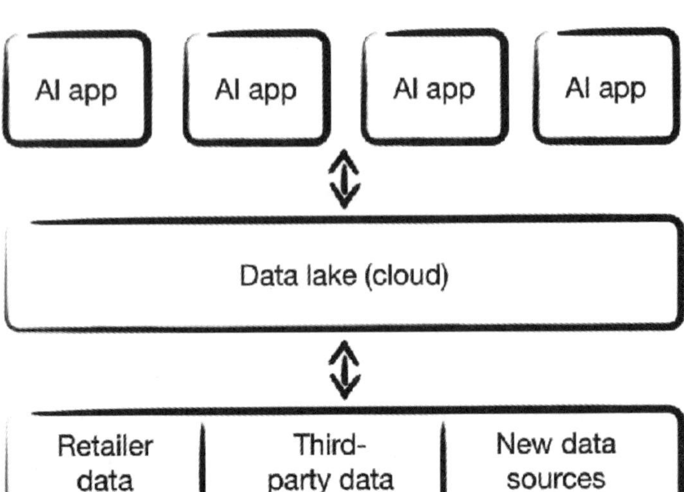

Walmart as a Platform

It's already happening at Walmart. Over the past several years, Walmart has been migrating more and more of its data to the cloud; in some cases Microsoft's Azure, but also Google's Cloud, and more recently, Walmart's own cloud. Walmart is quickly evolving into a platform business and developing a growing array of AI agents at its core that keep the business flowing.

The company has deployed an AI chatbot to automate vendor negotiations. Used so far for non-resale purchases, things like shopping carts, the technology has resulted in significant gains to efficiency. Surprisingly—or maybe not so surprisingly—the vendors actually prefer the new AI chatbot to dealing with human buyers. The chatbot responds nearly immediately, enabling the vendor to move quickly through the process rather than negotiating over months with a human being.

"Walmart has found that the chatbot closes deals in days rather than weeks and that nearly three out of four suppliers prefer negotiating with the chatbot. The technology enables Walmart to negotiate with 2,000 suppliers at once. Walmart's experience has been that 83 percent of suppliers find the chatbot easy to use, that the chatbot closes deals with 68 percent of the suppliers with which it's used and that it generates an average savings of 3 percent for the retailer."*

Walmart has for some time leveraged AI in supply chain capabilities like demand forecasting. The technology is spreading quickly into other operational areas as Walmart builds its digital doppelgänger.

Think about the growing number of massive Walmart distribution centers using Symbotic's automation. Only a handful of human beings needed to operate massive distribution centers. AI-powered automation doing nearly everything. The company is building localized eCommerce fulfillment centers using the Alert Innovation micro-fulfillment center tech it acquired. And Walmart is using a growing armada of self-driving trucks, moving orders between an automated online fulfillment center and the store where shoppers will pick the order up. The company's partnership with Gatik continues to grow, using Gatik's focus on short-haul, repeating route capabilities. Walmart is well along the path to becoming bionic.

Think about that: Online orders flowing in from shoppers, the

---

* PYMNTS, "Walmart Reportedly Finds 75% of Vendors Prefer Negotiating With Chatbot," April 26, 2023, https://www.pymnts.com/news/artificial-intelligence/2023/walmart-finds-75-percent-vendors-prefer-negotiating-with-chatbot/.

order automatically picked by a robotic fulfillment system, the order put into a self-driving truck for delivery to the store for customer pickup. A nearly fully automated end-to-end process involving very few human beings. Walmart as a retail digital platform.

AI at the Core

Platforms have AI at their core, AI agents making decisions, automating all the key processes needed to keep the business operating. And why not? Is making a financial loan decision any harder than making a decision to carry a new product? Or to price it? Or determine inventory requirements?

In all these examples, AI systems view the decision-making process as data science, leveraging data into insights, analytics, choices, and predictions that then automate operational processes and workflows.

"Information technology is no longer merely an enabler and optimizer of traditional processes and methods; instead, software makes up the actual operating core of the firm. Replacing traditional labor-and asset-intensive organizations, fueled by a pipeline of data and powered by algorithms, software constitutes the critical path in delivering value to the firm's customers. And because of these digital foundations, the firm is capable of generating increasing returns to scale, scope, and learning—and of overwhelming traditional business models."*

---

* Marco Iansiti and Karim Lakhani, *Competing in the Age of AI* (Boston: Harvard Business Review Press, 2020).

HuLoop Automation provides AI-powered, no-code intelligent automation, that is a good example of the types of AI applications that power up the core. HuLoop delivers digital transformation outcomes by integrating with existing technology already in place and tearing down organizational silos that have developed over time. It's designed to transform how work gets accomplished while freeing up human capacity to focus on other, more valuable work.

HuLoop offers a suite of AI-powered tools to automate a growing number of processes across retail industry operations. These are the types of AI applications ready to plug into digital operating platforms, transforming core operations across the enterprise. HuLoop's solution suite includes intelligent purchasing agents, merchandising agents, pricing agents, inventory agents, eCommerce and digital marketing agents, amongst others.

Marco Iansiti and Karim R. Lakhani are professors of business administration at Harvard Business School and authors of *Competing in the Age of AI: Strategy and Leadership When Algorithms and Networks Run the World*. They suggest the idea of an AI "factory."

"AI factories are at the core of the model, guiding the most critical processes and operating decisions, while humans are moved to the edge, off the critical path of value delivery. In its essence, the AI factory creates a virtuous cycle between user engagement, data collection, algorithm design, prediction, and improvement. It integrates data generated from multiple sources (internal or external to the firm) to refine and train a set of algorithms. These algorithms not only make predictions but also use the data to improve their own accuracy. The predictions then drive decisions and actions, either by informing human insights or by enabling an automated response. Hypotheses about changing customer behavior patterns, competitive responses, and process variations are tested through rigorous experimentation protocols that enable causal identification of changes that might improve the system. Data about usage and about the accuracy and impact of the prediction outcomes is then sent back into the system for further learning and predictions. And the cycle continually repeats."[*]

It's Not Just about AI

Artificial intelligence is being woven into many existing solutions and capabilities—sometimes to improve performance and other times as a marketing move given how hot AI is.

A report by Grocery Doppio, a data analytics firm, states that AI implementation in supermarkets is projected to grow by 400 percent within the next eighteen months, eliminating 18 percent of store associate positions, 73 percent of store tasks, and 53 percent of shopper queries.[†]

"These efforts are projected to generate $113 billion in value, with inventory management savings accounting for more than half

---

[*] Ibid.

[†] Timothy Inklebarger, "Use of AI in grocery stores to grow 400% by 2025," Winsight Grocery Business, June 22, 2023, https://www.winsightgrocerybusiness.com/technology/use-ai-grocery-stores-grow-400-2025?utm_source=Marketo&utm_medium=email&utm_campaign=NL_WGB_Today_06-26-23_09:00&LID=4020874&mkt_tok=NTYxLVpOU-C04OTcAAAGMlvn0FcfwAoJUVTS4eZYnJwXksgM26CM8WYC-tbht5_KGr5_YAXvgCUjEz87aV1Lh6i96F2fCsQjxTlD07-w5O0G3vdn66BYR-b0lQEUDKVqjtSQ.

at an estimated $58.4 billion, Grocery Doppio's study said. Advances in product assortment and pricing optimization through AI will save grocers $21.1 billion in merchandising costs, and another $15.7 billion is projected to be generated on marketing. AI will also save grocers $13.8 billion in store operation costs, $2.1 billion in customer service and $1.9 billion in IT/technology, according to the report."*

But here's the problem: Incorporating AI into existing solutions and capabilities can certainly deliver benefit *but it does not create a digital operating platform business model.* Just putting AI into existing capabilities maintains the siloed, disparate, and increasingly complex, structure that exists today. Yes, some benefit is created but this approach perpetuates the core problem: today's organizations and operations are increasingly misaligned with the exponential world of retail.

Platform Benefits

As Amazon has shown—and Walmart is experiencing—platforms are as much a business model as they are a technology foundation that's able to handle fast growth and scale while minimizing costs. Once built, platforms can handle additional volume with ease—again, ultimately the core of the business is automated using AI agents—just like Ant Financial. Pumping more volume through software is highly efficient and cost effective.

Amazon, again, provides an example of this. Tom Furphy was

---

* Ibid.

brought into Amazon in 2005 to help develop its CPG business. Amazon knew it wanted, and needed, to expand into grocery, given the sector's massive consumer penetration, size of the market, and frequency of purchase. That being said, Furphy and team faced some constraints. The first was Amazon's website experience, which was built to help shoppers quickly find one or two items and then tender a purchase. The other constraint was fulfillment; Amazon was using parcel delivery and many CPG products were either "breakable" (not conducive to shipping) or temperature sensitive.

Adding to the challenges was the fact that many of the CPG companies had a bitter taste remaining from Webvan's failure. Webvan was the dot-com company founded in 1996 with a vision to take grocery shopping online paired with home delivery. The company raised an estimated $1 billion between investors and an IPO, which it burned through before declaring bankruptcy in 2001.

Navigating the constraints while pursuing the goal of getting Amazon into the CPG business led to the development of Amazon's Subscribe & Save program. The idea was to find those products that had regular purchase frequency and unit economics (price) that supported providing a discount for signing up for a subscription and supported free shipping. According to Furphy, the effort began later in 2005 and Subscribe & Save was launched in 2007, after an eighteen-month period to build it. Subscribe & Save is on track to generate an estimated $40 billion in annual revenue in 2023.

Amazon's subscription program was a natural fit for its digital operating platform and business model. The program utilizes the data already captured through the company's online shopping site, it builds repeat, regular purchases, which, in turn, build loyalty and customer lifetime value for both Amazon and the participating CPG brands. More volume, more data, flowing through the platform at little incremental cost.

Walmart is already the largest grocery retailer in the United States with sales of over $300 billion; 60 percent of the company's U.S. sales are grocery, up from only 36 percent twenty years ago according to Scott Moses, partner at Solomon Partners. The company has over 4,700 stores across the United States and more than 90 percent of Americans live within 10 miles of a Walmart. Walmart is the largest online grocery seller with more than $40 billion in annual online grocery sales.*

Now, think about all that sitting atop a massive digital foundation built from enormous flows of data, the core of Walmart's business increasingly being handled by AI systems that continuously evolve and grow more effective through machine learning. Human beings moved increasingly to the edges of the business while more and more functions are handled through software automation. Walmart, like Amazon, is increasingly able to pump more and more transactions through the platform at negligible incremental cost. That's a platform as a business model.

What Does It Take?

So, what does it take to build a retail platform?

The first step is to think of retail in a new way. Yes, it's still the

* Scott Moses, 5 Things You Might Not Know," Solomon Partners, June 2023, https://solomonpartners.com/wp-content/uploads/2023/06/Scott-Moses-Solomon-Partners-5-Things-You-Might-Not-Know-Walmart-June-2023.pdf.

movement of product from manufacturers to distribution centers and from stores to shoppers. But it's envisioning the "mirror" of this physical world of retail as digital data and information flows.

That's where the magic happens.

But I believe for many retail executives the big challenge is thinking beyond physical warehouses, stores, products, and shoppers. The initial reaction from traditional executives to the platform idea is something like, "Well, digital platforms are all well and fine for companies like Netflix or Airbnb. They don't have to handle actual products like we do. You can't digitize a grocery supply chain and operation."

To understand and appreciate what companies like Amazon and Walmart are doing, you have to let go of how you've always seen the world of retail and *think differently*. As we discussed earlier on, results don't change until beliefs do. You have to believe that retail as a technology platform is possible if you want your organization to successfully navigate the shifting tides ahead.

And I think that's where the mirror image of the physical world of retail can be helpful. The mirror image is the digital data and information that is attached to each product and shopper flowing through your business.

The key to retail as a platform is *data*. And think about all the data there is across a retail business, especially companies like Amazon or Walmart. The ever-growing quantity and diversity of data demands a way to find what you're looking for. And that calls for a data catalog, akin to a library's catalog.

According to Oracle, "a data catalog is an organized inventory of data assets in the organization. It uses metadata to help organizations manage their data. It also helps data professionals collect, organize, access, and enrich metadata to support data discovery and governance.

When you go to a library and you need to find a book, you use their catalog to discover whether the book is there, which edition it is, where it's located, a description—everything you need so that you can decide whether you want it, and if you do, how to go and find it.

And now expand the power of that catalog to cover every library in the country. Imagine that you have just one interface and suddenly, you can find every single library in the country that has the copy of the book you're seeking, and you can find all the details you'd ever want on each one of those books.

That's what an enterprise data catalog does for all of your data. It gives you a single, overarching view and deeper visibility into all of your data, not just each data store at a time."* These tools provide a level of data visualization that would be difficult and incredibly time-consuming for data analysts to complete manually.

Platform Evolution

Building a retail platform takes some doing. It took Amazon three iterations as capabilities evolved and the business scaled so quickly. Thankfully, others can learn from the lessons provided by Amazon, Walmart, and others.

The process must begin with data, bringing all the data flowing

---

* Oracle, "What Is a Data Catalog and Why Do You Need One? accessed September 2023, https://www.oracle.com/au/big-data/data-catalog/what-is-a-data-catalog/.

through a retailer together in the cloud so it's available 24/7, organized via a data catalog, and easy to access via application programming interfaces (APIs). Marco Iansiti and Karim R. Lakhani are professors of business administration at Harvard Business School and authors of *Competing in the Age of AI: Strategy and Leadership When Algorithms and Networks Run the World*. They have extensively studied companies and organizations that have transformed to best leverage new AI capabilities, creating new business models in the process.

"Stage 1 is where organizations usually begin: with siloed data. We rarely see much in the way of barriers before the pilot stage (stage 2), because demonstrating the value of analytics-based decision-making can be done without big organizational and cultural shifts and is often done largely by vendors and consultants. But as we get to the data hub stage (stage 3), the organization must rearchitect itself to aggregate data from many siloed sources and use it to identify companywide opportunities. This is the point when substantial investment is needed and when the organization starts to understand that it will need to change. Not surprisingly, this is also when we see organizational resistance.

"Most important (and often most challenging) is the adoption of a clear, single source of truth to guide decisions on market opportunity, pricing, planning, and operational optimization. The consistent approach to data and analytics is most often associated with the creation of a centralized organization devoted to data sciences and analytics, frequently deployed across applications, products, and SBUs in a hub-and-spoke fashion. Although individual functions and product units inevitably ask for some flexibility to adopt unique capabilities and approaches, the data sciences team must not lose the ability to connect its organization."*

Chapter Summary

- Many retailers' systems and processes have developed into a byzantine collection of data, tech systems, and processes that becomes more inefficient as the company grows. These systems and structured need to be remolded to allow for the efficiency and adaptability that will serve retailers in the digital age.
- Incorporating AI into existing solutions and capabilities can certainly deliver benefit but it does not create a digital operating platform business model. Just putting AI into existing capabilities maintains the siloed, disparate, and increasingly complex, structure that exists today. Building a digital platform can allow retailers to handle additional volume with ease.
- Building a digital platform to serve your organization is an iterative process, and it will likely take several tries and countless tweaks to keep it moving. However, starting by bringing as many useful data points as possible together into an accessible interface is a great place to begin.

* Marco Iansiti and Karim Lakhani, *Competing in the Age of AI* (Boston: Harvard Business Review Press, 2020).

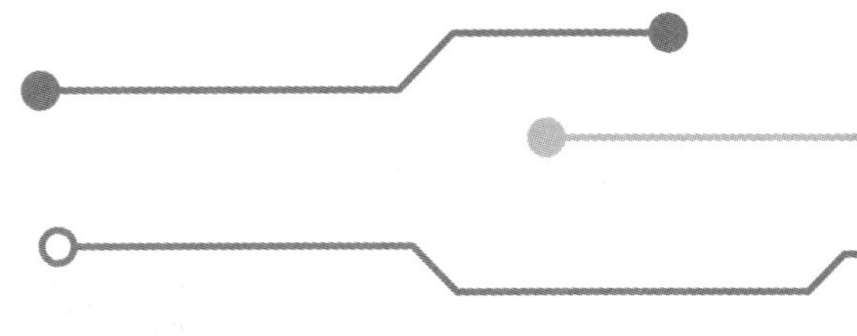

Chapter 12

PIPELINES AND PLATFORMS: CHANGING HOW VALUE IS CREATED

Traditional retail is a textbook example of a pipeline business model, value created by buying assets (products), controlling supply chains, and driving transactions through the store. Think of the flow as being more linear—goods moving from manufacturer, through the retailer's distribution centers, to stores, and then to shoppers.

As traditional operating businesses like retail scale, the growing complexity of integrating siloed systems and data, along with growth of human-based processes, begins to limit value creation. Economies of scale begin to plateau.

But now think about a digital operating platform business model. According to *Harvard Business Review*, "Platform businesses bring together producers and consumers in high-value exchanges. Their chief assets are information and interactions, which together are also the source of the value they create and their competitive advantage."*

Because core functions—in retail, think merchandising decisions, marketing, etc.—are handled by AI applications, added transactions pumped through the platform add little incremental cost. And each added transaction provides more data that feeds the machine-learning algorithms, that in turn improves the decision-making of the core AI applications. As volume scales, more data enables more effective decision-making by the AI apps, which in turn leads to growing sales from improved product assortment, more intelligent pricing, more effective marketing, and so on. This is just what Amazon has created.

---

* Marshall W. Van Alstyne, Geoffrey G. Parker, and Sangeet Paul Choudary, "Pipelines, Platforms, and the New Rules of Strategy," *Harvard Business Review*, April 2016, https://hbr.org/2016/04/pipelines-platforms-and-the-new-rules-of-strategy.

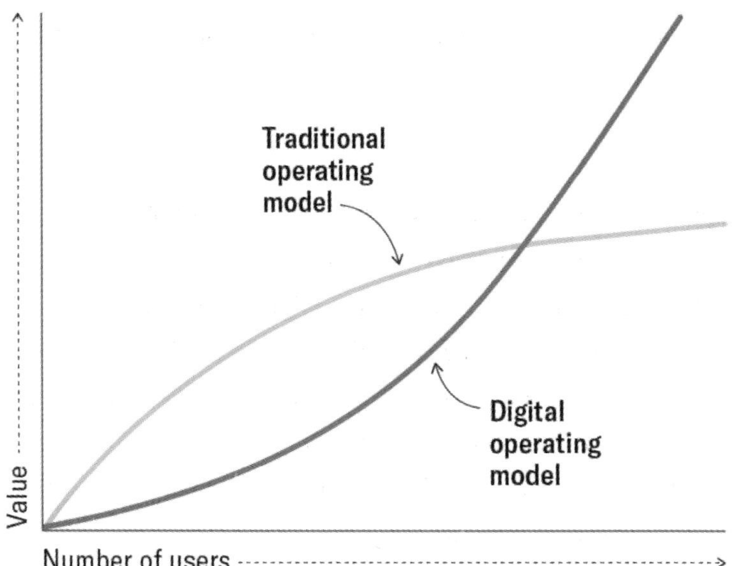

From: "Competing in the Age of AI," by Marco Iansiti and Karim R. Makhani. January-February 2020, Harvard Business Review.

"Ocado is an online grocery delivery company in the UK that redefined itself through its adoption of AI. Algorithms draw from the rich internal and external data collected from Ocado's purpose-built central platform to optimize all aspects of the service-delivery chain. From anticipating and averting potential shortages from the farms in their supply chain to real-time delivery route optimization, AI technologies drive the majority of the company's operations. Ocado also revolutionized its operations by setting up highly automated warehouse systems. In each of these warehouses—some as large as five city blocks—algorithms coordinate the movements of thousands of bots, miles of conveyors, and hundreds of thousands of grocery boxes to optimize efficiency and freshness.

Ocado is an AI company disguised as a supply chain company disguised as an online grocer."*

Development of Digital Platforms in Fast-Moving Consumer Goods (FMCG) Retail

In 2003, Kroger began working with dunnhumby to leverage the massive amount of shopper-identified transaction data the company was collecting through its loyalty program. Not long after, Kroger began mailing to shoppers targeted coupons, leveraging the insights gained from the data. Some of those targeted coupons were "rewards" giving the shopper savings on products they frequently purchase, and some of the coupons were designed to capture increased share of wallet from adjacent categories or products only purchased occasionally. This effort helped fuel what became a consecutive fifty-two-quarter run of same store sales growth. That run is unprecedented in FMCG retail, but it's what was happening beneath the surface that was extraordinary.

The following are excerpts from the 2010 Q3 earnings call transcript with comments provided by (at the time) Kroger CEO David Dillon and CFO Rodney McMullen:

> *"We were pleased with Kroger's third-quarter results and believe they demonstrate the continued success of our Customer First strategy. Increases in identical sales, tonnage, and loyal household count show our core grocery business is strong and resilient.*

---

* Digital, Data, and Design Institute at Harvard, "Competing in an AI-Driven World," September 2023, https://d3.harvard.edu/insights/competing-in-an-ai-driven-world/.

In the third quarter, the total number of households we served continued to grow. In fact, more customers across all segments from value to upscale visited our stores compared to a year ago. This tell us that Kroger's shopping experience and value proposition is striking a chord across a broad spectrum of shoppers.

While total household growth is important, we focused more attention and energy on increasing our number of loyal households. Kroger has been steadily increasing its number of loyal households for several years. In the third quarter, Kroger added loyal households at a faster rate than total households. The number of visits per loyal household remained about the same as a year ago, but the amount our loyal customers spend with us on a monthly basis continues to increase. This drives the solid identical sales results that Dave shared with you earlier.

We use this information to offer more compelling ads and weekly specials, and to offer the right product selection in our stores. Kroger has more insight in our customer behavior than most retailers today, and we use this data to reward them for shopping with us with special offers and coupon mailers."[*]

---

[*] Seeking Alpha, "Kroger Co. CEO Discusses Q3 2010-Earnings Call Transcript," December 2, 2020, https://seekingalpha.com/article/239801-kroger-co-ceo-discusses-q3-2010-earnings-call-transcript?open_reset_password=true&origin=confirm_registration&utm_campaign=%7Cconfirmation_link_registration&utm_medium=e-mail&utm_source=seeking_alpha.

As experienced practitioners of loyalty marketing in retail understand, higher-spending shoppers provide higher gross margins. It stands to reason: higher spending shoppers are purchasing more products from around the store, including more products at regular price. I can remember in data gleaned from our loyalty program seeing a 10 percent spread between the gross margin of more loyal shoppers vs. occasional or deal-seeking shoppers.

So, as Kroger began creating more higher-spending shoppers in its customer base—a growing portion of total sales coming from more profitable shoppers—the company's margins began to grow. Kroger, led by David Dillon as CEO throughout this time, wisely invested that increased margin back into more competitive pricing and customer services . . . both of which served to attract more new shoppers that were fed into Kroger's "machine."

What Kroger created, and is still running, is a virtuous cycle of value creation. Shopper data feeds precision marketing that in turn grows shopper spending and margin, that is in turn reinvested in competitive pricing and services to attract more shoppers . . . that provide more data, fueling the system.

Is this starting to sound familiar?

Now, think about what's happening at Kroger today. Kroger's 84.51° division is using cutting-edge data science, machine learning, and AI to automate that precision marketing. What used to take hundreds of data analysts, crunching through data, creating ever-finer audience segmentations to power up the targeted promotions, is today increasingly taken over by automated processes. In essence, Kroger is creating a digital platform fueled by shopper data that powers strategic precision marketing at the shopper level.

And that data also supports product assortment decisions, inventory decisions, pricing decisions, and more, all at a store level. And you can bet more and more of those decisions are being migrated to machine learning supported AI applications.

Tom Davenport is the President's Distinguished Professor of IT and Management of Babson College, a Digital Fellow at the MIT Initiative on the Digital Economy, and a senior advisor to Deloitte's Analytics and Cognitive practice. He wrote the following about Kroger's advanced use of machine learning in a *Forbes* article.

"The current approach to machine learning at 84.51° emerged from an initiative called 'Embedded Machine Learning' (EML). Scott Crawford leads the initiative, but it had multiple progenitors. 84.51°'s Chief Operations Officer, Milen Mahadevan, is a champion for automation of processes and products within the organization. 84.51°'s Shop, a custom-built BI platform that allows CPG customers to pull detailed reports about shopping behavior, is a successful example of BI automation. Shop evolved to replace and extend what had been ad hoc reporting. Embedding machine learning is the logical progression from ad hoc modeling/segmentation to automated processes that generate value through efficiency and improved accuracy.

"Many companies today are experimenting with machine learning, but 84.51° and Kroger have taken this AI approach to the next level. The 'Embedded Machine Learning' initiative, standardizing on an automated machine learning tool, and the three-stage machine learning methodology have all helped to create a 'machine learning machine.' Models are framed, developed, and deployed in the same way that a well-managed manufacturing organization might create physical products."*

Like Walmart, Kroger is transitioning to a digital operating platform business model. A growing number of core operating processes—decision-making—being handled by AI applications able to grow ever-more intelligent and effective as more and more transactions (data) are fed through the system.

Network Effects

Traditional retail has historically been a textbook example of a linear business, value created by buying assets, controlling supply chains, and driving transactions through the store. Digital companies though drive value creation through creating network effects. Traditional retailers have an opportunity to step onto the digital battlefield and access exponential value creation through the growth and expansion of their own digital networks.

Robert Metcalfe, co-inventor of Ethernet and co-founder of 3Com, a manufacturer of network cards for computers, was a keen observer of the power of networking.

"Metcalfe's law says that a network's value is proportional to the square of the number of nodes in the network. The end nodes can be computers, servers, and simply users. For example, if a network has 10 nodes, its inherent value is 100 (10×10=100). Add one more node, and the value is 121. Add another and the value jumps to 144.

---

* Tom Davenport, "84.51° Builds a Machine Learning Machine for Kroger," *Forbes*, April 2, 2018, https://www.forbes.com/sites/tomdavenport/2018/04/02/84-51-builds-a-machine-learning-machine-for-kroger/?sh=18b35d9064e1.

Non-linear, exponential, growth.

"Network effects have become an essential component of successful digital businesses. First, the Internet itself has become a facilitator for network effects. As it becomes less and less expensive to connect users on platforms, those able to attract them en mass become extremely valuable over time. Also, network effects facilitate scale. As digital businesses and platforms scale, they gain a competitive advantage, as they control more of a market. Third, network effects create a competitive advantage."*

Think of each digitally engaged customer as a digital node on a network. Now apply Metcalfe's Law of network value to retail. If a retailer has 1 million digitally engaged shoppers—digital nodes—the retailer's network has an inherent value of 1 million squared, or 1 trillion (1,000,000,000,000). Retailers, with their large customer base, have an inherent potential to create massive value if they are able to successfully engage with their shoppers across digital channels.

To realize that massive potential value, retailers must expand their digital network, fostering connections with other shoppers, other services, IoT devices, and more—inside and outside the store.

This view creates a need for new metrics. To play in this new world of digital value creation, retailers need to measure both their network size and the quality of customer digital engagement. Far too few C-level executives measure digital shopper engagement,

---

* Peter Fisk, "Metcalfe's Law Explains How the Value of Networks Grows Exponentially . . . Exploring The 'Network Effects' of Businesses like Apple, Facebook, Trulia and Uber," PeterFiske.com, February 17, 2020, https://www.peterfisk.com/2020/02/metcalfes-law-explains-how-the-value-of-networks-grow-exponentially-there-are-5-types-of-network-effects/.

knowing what percentage of their company's business is generated by those shoppers digitally engaged.

Pipelines and Platforms

So, network effects are vital to the new digital operating platform model. And interestingly, linear, physical asset-based pipeline models can be synergistic with the new digital network model. Just look at Apple. Apple's iPhone business is a traditional pipeline model. Apple creates and owns the physical product, manages the supply chain, and distributes the iPhones to end shoppers via stores and online. But combine the iPhone with Apple's App Store, which brings together app developers and consumers, and you have a platform.

"As Apple demonstrates, firms needn't be only a pipeline or a platform; they can be both. While plenty of pure pipeline businesses are still highly competitive, when platforms enter the same marketplace, the platforms virtually always win. That's why pipeline giants such as Walmart and Nike are scrambling to incorporate platforms into their models."*

Stop to think about retailers' infatuation with retail media networks (RMNs). Sure, those networks are generating billions of dollars in new, incremental revenue for the largest retailers. But those ads, nearly all of which are promoting products sold by the retailer, also generate additional product sales and transactions . . .

---

* Marshall W. Van Alstyne, Geoffrey G. Parker, and Sangeet Paul Choudary, "Pipelines, Platforms, and the New Rules of Strategy," *Harvard Business Review*, April 2016, https://hbr.org/2016/04/pipelines-platforms-and-the-new-rules-of-strategy.

feeding more transactions through the digital operating platform. And all the data gathered through those campaigns—who was targeted with what offer, which shoppers clicked on the ad, who of them visited the store and made a purchase—all feeds into the machine-learning algorithms to improve the network's performance.

And do you suppose all those third-party sellers want to advertise their products to Walmart shoppers? More revenue for Walmart Connect, the company's RMN, and more transactions flowing through the digital operating platform. As Walmart's business—both in physical stores and online—continues to accelerate, it enjoys accelerating returns due to its developing digital operating platform.

Prime Gets Personal

Amazon's annual Prime Day is another example of leveraging the digital operating platform to accelerate growth. Amazon's Prime Day has evolved into an annual merchandising event that has driven competitors like Walmart, Best Buy, Target, and others, to launch competing sales. Prime Day debuted in 2015, in which Amazon shoppers purchased 34 million products; in 2023 that number exploded to 375 million items.* To continue that kind of growth, Amazon is leveraging its data, digital operating platform, and AI, to make Prime personal.

"By leveraging machine learning, tweaking existing features, and introducing a few new ones, Amazon aims to put the deals that are most likely to appeal to you directly in your line of sight. It's not the first year Prime Day has been personalized (that would have been in 2021—and engagement shot up 400 percent), but it's likely going to be the most noticeable.

"To make the site hyper-personalized for Prime members, Amazon is using AI and deep learning, says Nestares Pleguezuelo, looking at everything from shipping and browser histories to wish list. While there's not necessarily anything new about that, the Prime Day implementation of it will be to a higher degree than you've seen before. 'I think it's always been in the DNA [of the company],' she adds. 'But bringing it forward . . . is very powerful. Our ambition over time is to make the merchandising of our site more personalized.'"†

---

* Amazon Staff, "The First Day of Prime Day Was The Single Largest Sales Day Ever on Amazon, Helping Make This the Biggest Prime Day Event Ever," Amazon, July 13, 2023, https://www.aboutamazon.com/news/retail/amazon-prime-day-2023-stats#:~:text=Over%20the%20course%20of%20the,biggest%20Prime%20Day%20event%20ever.

† Chris Morris, "Amazon's Prime Day 2023 Will Be All About Personalization," Fast Company, July 10, 2023, https://www.fastcompany.com/90919743/amazon-prime-day-2023-will-be-all-about-personalization.

Amazon: Data-Driven Disruption

Amazon sells more than 12 million products itself, and more than 600 million products if counting Amazon Marketplace sellers.* Nine out of ten consumers price-check a product on Amazon. And don't forget about Whole Foods, where Prime members are identified to their transactions in return for deals. The company's knowledge of product purchases and purchase intent across countless categories is itself a massive trove of incredibly valuable data.

But add to that Amazon's world-class supply chain extending to a home delivery network soon to rival UPS and FedEx. There are an estimated 70 million† Alexa users in the United States—imagine the data and insights Amazon gathers as users interact with the system. There are more than 160 million‡ Prime members in the United States; consider the information gained as those millions of people use Prime Video and Prime Music. And as Amazon scales deployments of its Just Walk Out computer vision tech and smart carts in a growing number of stores, the company gains knowledge of each individual's behavior inside the store.

And there's more. Amazon is in the pharmacy business, providing

---

* AMZ Scout, "Amazon Statistics: Key Numbers and Fun Facts," accessed September 2023, https://amzscout.net/blog/amazon-statistics/.

† Harry Guinness, "Alexa, Why Are You Losing So Much Money?" Popular Science, November 23, 2022, https://www.popsci.com/technology/amazon-alexa-lose-money/.

‡ Statista, "Number of Amazon Prime Users in the United States from 2017 to 2022 with a Forecast for 2023 and 2024," accessed September 2023, https://www.statista.com/statistics/504687/number-of-amazon-prime-subscription-households-usa/.

prescription drugs. They recently expanded into healthcare through the acquisition of One Medical. Amazon company knows the size and layout of your home if you use a Roomba floor cleaner. Amazon knows your web activity if you use their Eero networking equipment. Amazon is expanding its payment services online and across physical stores with new services like Amazon One, using a scan of your palm to authorize payment.

I would suggest that Amazon has a more comprehensive, 360-degree view of the consumer than any company in history as it is able to connect all those disparate data points. All that data—much of it tied back to the shopper—flowing into Amazon's digital operating platform, available across the enterprise to develop new services, new products, and to use in improving marketing relevancy. All that data feeding Amazon's machine-learning algorithms, improving the capability of its AI applications.

To be fair, it appears that Amazon is deliberately separating data across product lines to protect the integrity of customer relationships. But it is also probably fair to assume that as services begin to converge—like food retail and healthcare—that the data will become more consolidated. Regardless, Amazon's breadth of data makes Kroger's industry-leading depth of knowledge on the company's 60 million households seem like small change.

Chapter Summary

- As volume scales, more data enables more effective decision-making by the AI apps, which in turn leads to growing sales from improved product assortment, more intelligent pricing, more effective marketing, etc.

- Digital companies drive value creation through creating network effects. Traditional retailers have an opportunity to step onto the digital battlefield and access exponential value creation through the growth and expansion of their own digital networks.
- Linear, physical asset-based pipeline models can be synergistic with the new digital network model, so traditional retailers don't have to abandon all of their age-old processes and approaches to adopt a digital mindset—they can work together.

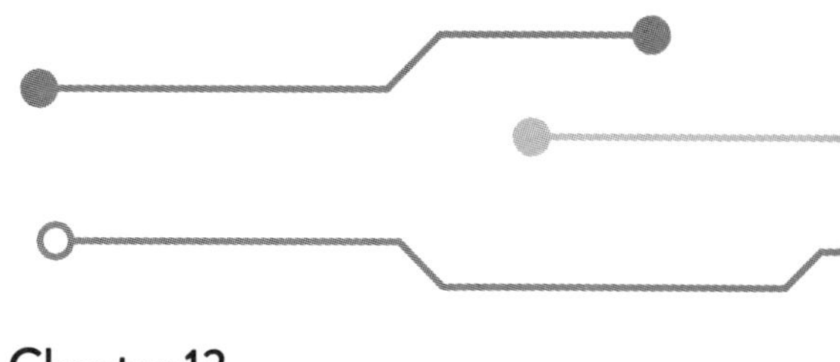

Chapter 13

BIONIC RETAIL

"Steve Austin, astronaut. A man barely alive."
"Gentlemen, we can rebuild him. We have the technology. We have the capability to make the world's first bionic man. Steve Austin will be that man. Better than he was before. Better—stronger—faster."
—Oscar Goldman, The Six Million Dollar Man

The Six Million Dollar Man was a television series popular in the 1970s. It tells the story of Steve Austin, former astronaut and now a test pilot, who crashes and is rebuilt with bionic capability. In the context of this book, we could say that Oscar Goldman had an inspired vision: To take a "broken" man and rebuild him, using advanced technologies to replace or augment capabilities that had been lost or damaged. To make him better, stronger, and faster.

Fifty years after *The Six Million Dollar Man* debuted, in a case of fiction becoming reality, technology is bringing Steve Austin's capabilities to life.

The Science Corp., based in Alameda, California, has developed a prosthetic it calls Science Eye. In early testing, "once it's been proved safe and effective, it'll be implanted on top of, and inside, the eyeballs of human patients suffering from diseases where the eye's light-sensing cells have died. The idea is to coax other cells within the eye to receive and translate light signals."*

A team at the University of Utah has developed a bionic leg that weighs only 6 pounds, about half what a standard prosthetic leg weighs. "What sets this prosthetic apart is that it uses motors, processors, and advanced artificial intelligence that all work together. It gives amputees more power to walk, stand up, sit down, and navigate stairs and ramps according to a news release. The extra power from the prosthesis makes these activities easier and less stressful for amputees, who normally need to over-use their upper body and intact leg to compensate for the lack of assistance from their prescribed prosthetics."†

A team at the Center for Bionics and Pain Research in Sweden has successfully tested a bionic arm and hand that allow the user to control each finger. Rewiring the nerves in the amputated limb "allows users to generate more complex movements with the bionic hand, such as flexing and extending all five fingers to pick up small

---

* Jackson Ryan, "The Bionic Eye That Could Restore Vision (and Put Humans in the Matrix)," CNET, May 28, 2023, https://www.cnet.com/science/biology/features/the-bionic-eye-that-could-restore-vision-and-put-humans-in-the-matrix/.

† Tamara Vaifanua, "The U Unveils World's Most Advanced Bionic Leg," October 5, 2022, https://ksltv.com/507776/the-u-unveils-worlds-most-advanced-bionic-leg/#:~:text=The Utah Bionic leg is,intelligence that all work together.

objects or type on a keyboard. Then we put electrodes inside them so we can extract that information, use an artificial intelligence algorithm to look at those electrical pulses that come from all the different sources and learn what is the patient trying to do. And once it's learned, which happens very quickly, then you can tell the prosthesis what to do."*

The cochlear implant, which uses very small electrodes as bionic replacements for hair cells within the ear, was developed in the 1960s to help restore hearing. "Yet, beside incremental improvements to the implant's algorithms and coding systems over forty years of operation, the underlying technology has largely remained unchanged."† A human clinical trial is underway in Australia to use gene augmentation therapy to improve hearing outcomes for people using cochlear implants.

But we're just getting started. The Living Bionics‡ project is funded by the European Research Council and is focused on better integrating bionic devices like cochlear implants and bionic eyes with the nervous system. MIT has developed a procedure that protects the nerves and muscles of an amputated limb so the new

---

* Camille Bello and Roselyne Min, "New Bionic Hand Allows Users to Control Each Finger with Unprecedented Accuracy," EuroNews.Next, July 14, 2023, https://www.euronews.com/next/2023/07/14/new-bionic-hand-allows-users-to-control-each-finger-with-unprecedented-accuracy.

† Heidi Vella, "Fusing Biology with the Bionic Ear," Engineering and Technology, July 7, 2023, https://eandt.theiet.org/content/articles/2023/07/fusing-biology-with-the-bionic-ear/.

‡ CORDIS, "Living Bioelectronics: Bridging the Interface between Devices and Tissues," European Commission, accessed September 2023, https://cordis.europa.eu/project/id/771985.

artificial limb can continue to communicate with the brain.*

This kind of work sits alongside more direct neural interfaces like Elon Musk's Neuralink, which is focused on a human-technology hybrid future: "Create a generalized brain interface to restore autonomy to those with unmet medical needs today and unlock human potential tomorrow."†

And Musk is not the only one working in this space. Synchron, mentioned earlier in the book, is also developing implants to aid a direct brain-computer interface. A Google search finds more than a half-dozen companies actively working to bring brain-computer interfaces to market.

Ray Kurzweil firmly believes technology will change us . . . literally. "The merger [between biological and machine intelligence] has already started. I mean, first of all, this [his cell phone] is not inside my body and brain, but it's pretty close. But there are people with computers in their bodies and brains [already]. There are computerized artificial pancreases that act just like the real organ."

"We have three profound overlapping revolutions, sometimes called GNR: G for genetics, which is biotechnology; N for nanotechnology, which is basically reprogramming matter and energy at the level of molecules, using information processes; R stands for robotics, but it really refers to artificial intelligence—creating intelligent machines whether they're robots or not. These are interacting in many ways. Ultimately, every industry is certainly affected. Those

---

* CBS Boston, "Brain-Controlled Bionic Limbs Developed At MIT," December 17, 2018, https://www.cbsnews.com/boston/news/mind-controlled-bionic-limbs-developed-at-mit-brigham-womens-hospital/.

† Neuralink, accessed September 2023, https://neuralink.com/.

industries that actually become information technologies go from linear progress—1, 2, 3, 4, 5—to exponential progress—2, 4, 8, 16—and that's a very profound change."*

Growing physical automation of traditional jobs is one thing. But like Neuralink's brain-computer interface, new AI tools are being integrated into the brain of retail and carrying out a growing number of functions. What's next is linking retail's "brain" to the rest of the organization.

Nervous System

In the human body, the nervous system serves as the "connector" between the brain and the rest of the body. "The nervous system is made up of all the nerve cells in your body," according to the National Library of Medicine. "It is through the nervous system that we communicate with the outside world and, at the same time, many mechanisms inside our body are controlled. The nervous system takes in information through our senses, processes the information, and triggers reactions, such as making your muscles move or causing you to feel pain. For example, if you touch a hot plate, you reflexively pull back your hand and your nerves simultaneously send pain signals to your brain. Metabolic processes are also controlled by the

---

* McKinsey & Company, "Strategic Practice: IT Growth and Global Change: A Conversation with Ray Kurzweil," *McKinsey Quarterly*, January 2022, https://www.mckinsey.com/~/media/McKinsey/Business Functions/McKinsey Digital/Our Insights/IT growth and global change A conversation with Ray Kurzweil/IT growth and global change A conversation with Ray Kurzweil.pdf.

nervous system."*

Traditional retail has a nervous system, but it's largely composed of people, working away in their siloed activities, communicating as needed. Someone in procurement emailing the marketing department to tell them the product featured in the upcoming ad will not arrive in time. The store manager noticing an issue with the frozen food case and calling maintenance. Someone in merchandising noticing that a competitor dropped a price on a key item and messaging someone in pricing to make an adjustment.

Unlike our biological nervous system, the signals flowing across the traditional retail organization are analog, most times a person messaging another person. By default, the system is slow and inefficient, sometimes fraught with errors. While traditional retail's "nervous system" has worked for decades, it is increasingly archaic when going up against an AI-powered digital operating platform.

Digital Nervous System

Imagine the nervous system of a digital retailer. Digital "nerves" send electrical (digital) signals to the brain, where they are processed, and responses are sent back out across the network. The AI system monitoring sales data notices a decline in the number of "gold" shoppers, triggering an automated personalized campaign to regain them. The AI monitoring the supply chain sees that a shipment of a key item is going to be significantly delayed, and it triggers a message to the buyer to review options. The automated distribution center messages

---

* National Library of Medicine, "How Does the Nervous System Work?" National Center for Biotechnology Information, October 28, 2009, https://www.ncbi.nlm.nih.gov/books/NBK279390/.

the receiver in store #32 that the delivery truck has left and will arrive in twenty minutes. All of this activity happens automatically, digital signals flow across the enterprise, increasingly in real time.

The digital retail nervous system is complex. Data that are the equivalent of red blood cells in the human body, flow throughout the enterprise. And, like real blood, inaccurate data, which are the equivalent of cancerous cells, can corrupt the integrity of the whole operation. The data being processed—consciously by humans in the loop, and unconsciously by AI and machine-learning applications (in real time 24x7x365). Decisions, again, either made by humans or automatically by AI agents, are sent back across the enterprise, executing the directed action.

The response to those decisions is recorded, creating automated feedback loops using AI and machine learning to make the system self-learning. Does something like a digital nervous system exist? The answer is yes, and we need to look no further than Palantir Technologies to understand what this looks like.

Powering the Digital Nervous System

Palantir Technologies was founded in 2003 by Peter Thiel and Alex Karp, along with a few others from the PayPal "mafia." Thiel and team were motivated by the events of September 11, 2001. Thiel believed that, while the U.S. intelligence community was very good at collecting data, they were challenged to turn that data into actionable insights. Thiel leveraged the approach PayPal had taken in developing AI and machine-learning capabilities to control fraud to develop capabilities to help the intelligence community.

Over time, Palantir expanded beyond working with the intelligence community in the United States and a handful of other friendly governments, to bring its data and technology prowess to bear in other areas with massive data sets, such as financial services and consumer goods retail. The company was actively involved in monitoring and projecting the spread of Covid during the pandemic and is active in the war in Ukraine, with their technology assisting the government with overseeing the logistics to resettle Ukrainian refugees and contributing to strategic military operations. One of their claims to fame was working with the U.S. intelligence services to direct Seal Team 6 to Bin Laden's final hiding place.

The core of Palantir's capabilities rests on Foundry, the company's operating system.

> *Foundry is a highly available, continuously updated, fully managed operating system that supports applications spanning from cloud hosting and data integration to flexible analytics, visualization, model building, operational decision-making, and decision capture.*
>
> *Having all of these capabilities available as part of a unified platform helps protect against the friction and risks associated with siloed technologies, and ensures a seamless experience where data history, security, and privacy are protected and maintained.*

As detailed in the following graphic, there are four basic steps to deploy and use Palantir's "digital nervous system" platform.

1. Palantir creates a bidirectional integration with a company's data and existing key systems like ERP systems, CRM, etc.

2. Next is creating a digital twin, the ontology, of the company's business including products, customers, and operations.
3. Foundry provides an intuitive, graphical representation of the complex decision processes that are part of a retailer's operations.
4. The ontology supports common language queries for decision-making, scenario planning, analytics, and more.

Palantir believes in having human beings in the decision-making process, knowing that a human may have experiential, contextual, even real-time knowledge of a situation that would impact a decision. What makes Palantir's approach so powerful is that each human-based decision serves to feed the machine learning, making the system smarter over time.

Palantir develops AI and machine-learning applications to address a company's specific operational challenges and ongoing operations, things like demand forecasting, promotion scheduling, pricing, and more. And those applications continuously learn through feedback loops and the human-in-the-loop decisions. Over time, those applications become smarter and smarter, requiring less and less human interaction. Decisions, either human directed or automated through AI, are then executed back through Palantir's tech stack.

It is this tightly woven approach to data, AI, machine learning, insights, and decision execution that positions a company like Palantir to become the digital nervous system for retail.

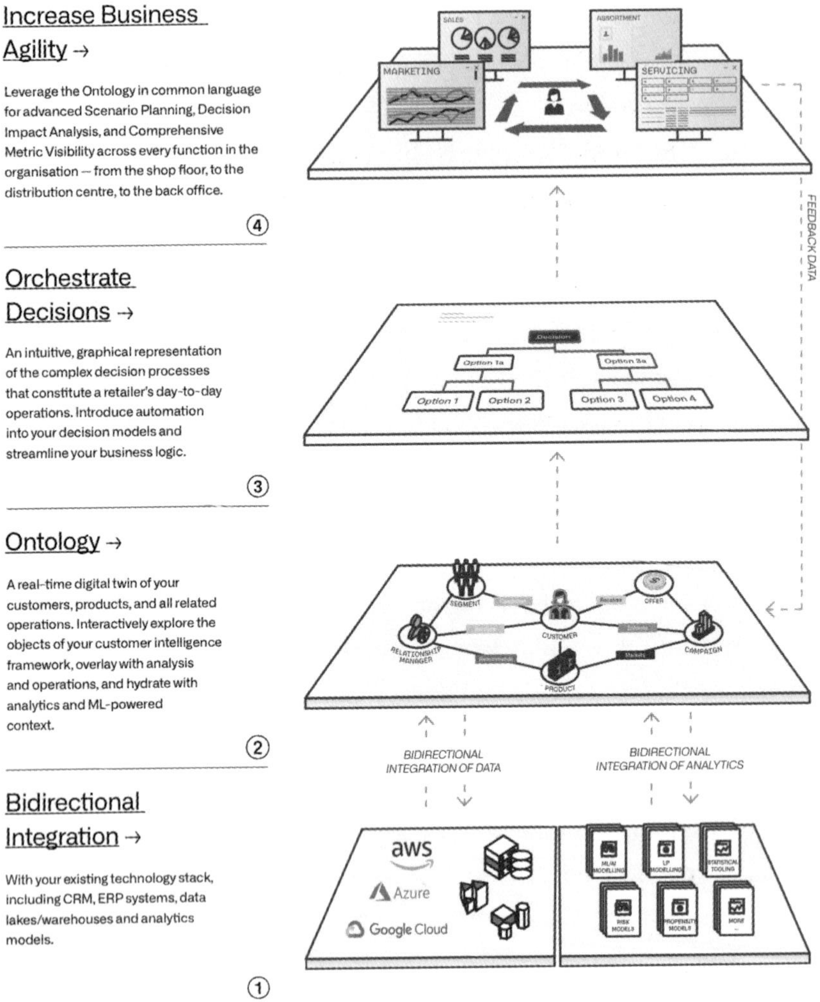

Increase Business Agility →
Leverage the Ontology in common language for advanced Scenario Planning, Decision Impact Analysis, and Comprehensive Metric Visibility across every function in the organisation – from the shop floor, to the distribution centre, to the back office.
④
Orchestrate Decisions →
An intuitive, graphical representation of the complex decision processes that constitute a retailer's day-to-day operations. Introduce automation into your decision models and streamline your business logic.
③
Ontology →
A real-time digital twin of your customers, products, and all related operations. Interactively explore the objects of your customer intelligence framework, overlay with analysis and operations, and hydrate with analytics and ML-powered context.
②
Bidirectional Integration →
With your existing technology stack, including CRM, ERP systems, data lakes/warehouses and analytics models.
①
SALES
ASSORTMENT
MARKETING
SERVICING
FEEDBACK DATA
Option 1a
Option 3a
Option 1
Option 2
Option 3
Option 4
SEGMENT
OFFER
CUSTOMER
RELATIONSHIP MANAGER
CAMPAIGN
PRODUCT
BIDIRECTIONAL INTEGRATION OF DATA
BIDIRECTIONAL INTEGRATION OF ANALYTICS
aws
Azure
Google Cloud
ML/AI MODELLING
LP MODELLING
STATISTICAL TOOLING
RISK MODELS
PROPENSITY MODELS
MORE

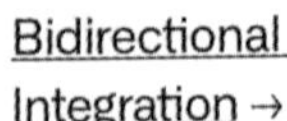

A New, Emerging Retail Species: Bionic Retail™

Merriam-Webster defines bionic as "having normal biological capability or performance enhanced by electronic or electromechanical devices." As we look about the industry, we can see the emergence of a new retail species—the bionic retailer.

Look across the retail enterprise today and you'll find a mosaic of automation deployed. What more are Walmart's hyper-automated distribution centers than "normal biological capability enhanced by electromechanical devices"? Symbotic's automation technology enhancing and replacing what has been biological (human) capability—slotting pallets, picking orders, loading trucks—with artificially intelligent robots. In a sense, this is the retail equivalent of Steve Austin's bionic arm.

More and more food production is being handled by robotics, everything from bread baking to salad making to Chipotle's avocado prep robot designed to increase production of guacamole; in reality these are bionic replacements of what was human activity.

Computer vision systems supporting cashier-free shopping like Amazon's Just Walk Out, or Focal AI's computer vision shelf monitoring, or Simbe's shelf-scanning robots, are the retail equivalent of Austin's bionic eye.

The retail industry is transforming right before our eyes, and yet many do not understand what is emerging. Retail executives are heads-down, focused on the tactical demands of the dynamic retail industry—the day-to-day issues they've been dealing with for years. And many are doing good things as they work to keep up, deploying automation in distribution centers, AI-powered marketing capabilities, micro fulfillment centers, and much more. But few have the time to step back to fully realize what's being created.

As all those disparate digital systems and capabilities deployed across the enterprise are connected through a digital nervous system, something new—a new retail species—is emerging: Bionic Retail.™

But the secret to Bionic Retail™ is the digital nervous system connecting all the pieces. An automated distribution center alone is like Steve Austin's bionic leg disconnected from his nervous system, it may function and add benefit, but it is when it's connected that things come alive.

Think of Bionic Retail™ as an ecosystem on steroids. As many Apple users know, the synergy Apple has created across its products is powerful, my MacBook Pro connected to my iPad, that's connected to my iPhone, and my Apple Watch. A notification sent to me viewable on any or all devices simultaneously. Core applications and data synced across devices, making me more productive and efficient.

The power of Bionic Retail™ lies in the digital nervous system that not only "senses" what's happening across a retailer's extensive operations but enables automated AI or humans in the loop to execute decisions across the organization based of these observations. An example of this is a signal from the Replenium "brain" to the automated micro fulfillment center to add three items to the order for the Hawkins household that are due for replenishment.

Or an AI-like Birdzi able to automatically "sense" a change in a shopper's behavior, generate shopper-specific promotions, and communicate them to the shopper—all without human intervention. The system recording the shopper's response, or lack thereof, to feed back into the machine learning to improve the initiative for next time.

Imagine the store clerk, wearing Augmodo's AR-powered smart

glasses, fulfilling eCommerce orders early in the morning, and the glasses automatically detecting extensive out-of-stocks for Coca-Cola products. Without the clerk doing any administrative work other than walking the store, the system automatically pings the local distributor to prioritize routing a delivery truck to the store to replenish the shelves.

Or the manager of store #46 learning that a new competitor will be opening just down the street and notifying the head office. That knowledge initiates a marketing campaign—again, executed through the digital systems—to retain shoppers, along with providing that competitive intelligence to the demand forecasting system to make adjustments in ordering. All that information and all those decisions—both human- and AI-driven—captured to inform machine learning to improve future results.

Bionic Cognition

Life on Earth developed over millions of years, rising from the primordial soup. Single-cell organisms gave way to more complex life, and eventually that life rose from the sea to inhabit the land. For nearly two million years, multiple human-like species co-existed, none gaining any lasting advantage. But then, as Yuval Harari, in his history-spanning book *Sapiens* explains, around 70,000 years ago a genetic mutation occurred, triggering a cognitive revolution that enabled Homo sapiens to quickly advance beyond all other species.

That cognitive spark changed the world, enabling imagination, conceptual thinking, expanded language, and more, giving this new sapient species the ability to rule the world. That first spark of cognitive intelligence, in a relative blink of an eye, raised Homo sapiens to a new plane of existence.

The retail industry is approaching a comparable "cognition event," when the disparate technological capabilities being deployed across today's retail enterprises are integrated—and connected—to a digital nervous system. As this happens, the Bionic Retail™ organization will quickly advance beyond slower, less intelligent, retail species. And this retail evolution is already underway, as companies like Walmart and Kroger accelerate up the technological growth curve on their way to becoming bionic.

"Walmart is developing its own artificial intelligence language models to improve supply chain efficiency and better connect with customers," CEO Doug McMillon shared.

"As it relates to technology, our approach to new tools like generative AI is to focus on making shopping easier and more convenient for our customers and members, and helping our associates enjoy more satisfying and productive work," McMillon said. "Ultimately, the power of generative AI or any technology is only as good as the data that powers it. Our data assets are unique, and we're excited about the potential to leverage them in new and impactful ways."

The language models Walmart is developing will "unlock value for shareholders through the combination of our physical automation work with our data and increasingly intelligent software," McMillon added.*

---

* Timothy Inklebarger, "Walmart CEO 'Excited about What's Possible' with AI," Winsight Grocery Business, August 17, 2023, https://www.winsightgrocerybusiness.com/technology/walmart-ceo-excited-about-whats-possible-ai

Bionic Benefits

Bionic Retail™ makes the business more efficient, as the benefits of both physical and business process automation are realized. But Bionic Retail™ also enables the retailer to grow and solidify customer relationships at scale. And to me, that's the enormous potential technology provides.

Any digital shopper facing engagement—whether it is an email, a website, an app, or a digital sign or kiosk in the store—provides the opportunity to make that interaction relevant to that individual shopper in that moment in time and location. The technologies to make this happen are now in the marketplace, tools using AI and machine learning, to automate personalization at massive scale.

The only reason a retailer would not deploy these types of capabilities is either ignorance of the available solutions or lethargy. It's like the retailers that have simply created a digital version of the mass promotion filled weekly ad rather than use technology to make that ad relevant to each individual shopper.

So, the promise of Bionic Retail™ is three-fold:

1. To make retail **better**—AI making more effective decisions.
2. To make retail **faster**—automation of both physical and business processes.
3. To make retail **stronger**—more resilient and powerful customer relationships.

Beware Inflammatory Responses

When devices like cochlear implants or brain–machine interfaces are implanted in the body, they trigger an inflammatory response.

Many times, the metals used in devices like these are recognized as foreign by the human nervous system, and the implants are either rejected by the body or tolerated rather than integrated, scar tissue walling off the device, limiting the intended benefit. If not done correctly and intentionally, transitioning your organization from traditional to bionic can have the same effect.

Becoming a bionic retailer is not just about plugging in technology. Just as in human beings, technology must be carefully integrated and synthesized into how the organization thinks and reacts. For the retailer wishing to become bionic, this means using many of the tools discussed throughout Part 2 to develop the organizational structure and culture to suppress the "immune response" and fully integrate new technologies into the organization.

The work to be done here should not be underestimated. As the world accelerates around us, being open to new possibilities and understanding that beliefs may change are necessary for adapting to a fast-changing environment. How retailers bring their associates along on the journey into the future will increasingly dictate success or failure.

Of most importance, though, is having a methodology to guide you and your organization into the future. A process for regularly identifying beliefs and practices that inhibit change and the successful implementation of new innovation to create the future. A process to develop tech-first thinking and strategy. And a process for discovering and understanding new innovation flowing into the massive retail industry.

In our experience there are really three things to focus on:

1. Developing an inspired vision.
2. Committing to action.
3. Developing a doctrine to guide the business into the future and instill it in the organization.

Don't Crash

The emergence of the bionic retailer is underway. And those retailers that are too slow to act will be left behind. The IHL Group calls out just how much top retailers will benefit from AI, especially as these large companies have the resources to build out their own data models and begin building, testing, and deploying these processes without disrupting their operations, which is something that is very difficult for small to midsize retailers. The analyst firm projects that Amazon and Walmart alone will realize an additional financial impact of more than $580 billion through 2029. Kroger is also in the top five retailers expected to gain significant benefit. Slower-moving retailers will be at an increasing disadvantage, as leaders exploit the revenue gains and cost savings made available through tech innovation.

I'll go further and say that many traditional retailers are like Steve Austin moments before the crash, thinking everything is going great and headed in for a picture-perfect landing. As said earlier in the book, traditional retailers are riding high, coming off pandemic-powered sales and inflation-boosted margins. But as we pass the inflection point on technology's exponential growth curve, a swarm of innovation butterflies are about to be sucked into the jet engines and it's not going to be pretty. Flameout is imminent.

Chapter Summary

- Bionic Retail™ is built on a unified data foundation. The retail industry continues to be challenged by maintaining accurate product data, and this needs to be addressed moving forward, especially as product related attributes continue to explode.
- Breaking down functional data and application silos makes it easier to deploy new AI and machine learning applications across core business processes. This in addition to deploying new automation of physical processes.
- A digital nervous system brings Bionic Retail™ to life, connecting all the systems across the enterprise together. That connectivity powers increased efficiency and effectiveness.

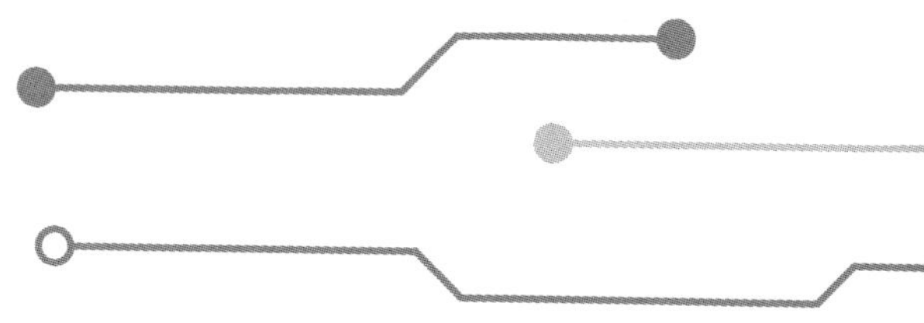

Part 4

PREPARING FOR A WORLD ARRIVING SOONER THAN EXPECTED

"Make it so."

— Captain Jean-Luc Picard

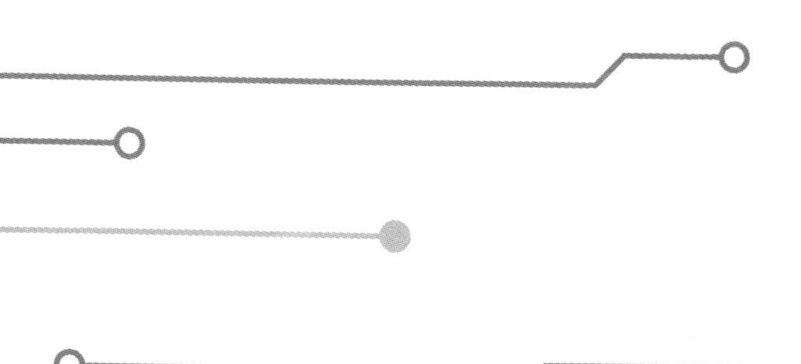

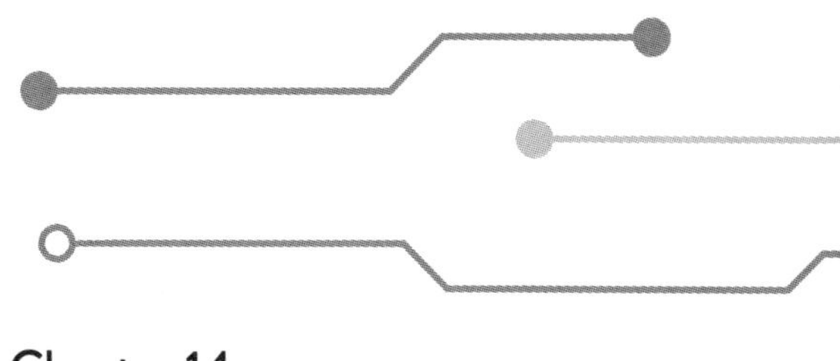

Chapter 14

EMERGING BUTTERFLIES

The Great Alaska Earthquake, also known as the Good Friday Earthquake, occurred on March 27, 1964, and left an indelible mark on the landscape of Alaska. With a magnitude of 9.2, it remains the most powerful earthquake ever recorded in North America and the second-largest earthquake in recorded history. The seismic event not only caused widespread destruction but also dramatically altered the physical geography of the region.

One of the most striking changes brought about by the Great Alaska Earthquake was the vertical displacement of the land. Along the coastline, large areas were uplifted or subsided, leading to significant rearrangement of the landscape. The Prince William Sound region experienced uplifts of up to 9 meters (30 feet), while areas along the Turnagain Arm subsided by as much as 2.4 meters (8 feet). This sudden vertical movement resulted in the emergence of new land and the flooding of previously dry areas, forever changing the configuration of the coast.

Landslides were another consequence of the earthquake,

significantly transforming the physical features of the affected areas. The intense shaking destabilized slopes, triggering mass movements of soil, rock, and debris. Landslides occurred on mountainsides, hillsides, and coastal cliffs, generating immense volumes of material that covered large areas. Some landslides dammed rivers, creating temporary lakes, while others blocked roads and disrupted transportation networks. The debris from these landslides forever changed the contours of the land, leaving scars and deposits that stand as lasting reminders of the earthquake's impact.

Many of the innovations flooding into retail are variations on a theme. There are multiple e-commerce solutions, each with some different features but all are basically the same. Or we see artificial intelligence (AI) being added to existing capabilities, improving an already established product.

But like the Great Alaska Earthquake, there are several emerging technologies that are poised to create fundamental change in the retail landscape. These are the massively disruptive butterflies emerging that promise to dramatically alter how we engage and interact with the world around us.

This is important to keep front of mind: as you read about these trends, don't evaluate them by what's possible today. Each of these is being powered by exponentially growing technologies—things will happen much faster than you think. Remember, the difference between linear thinking vs. exponential thinking as well as the destination either will get you to.

These technologies fundamentally change the way we engage with the world around us.

Disruption

When I was diagnosed with osteoarthritis in my hip, the doctor was ready to get me right on the schedule for a hip replacement within just a couple days. Not one to rush into surgery—once they start cutting it's tough to go back—I put him off, wanting to do some research first, which is just what I did.

I studied the different kinds of implants and surgical methods. And while hip replacements have advanced tremendously in the past twenty years, there are still complications in a small number of cases. One of the things I worried about was being able to ski—I have been skiing since I was five years old, and now I live in Colorado, so I wasn't ready to give it up.

I also began to research alternative treatments. While there are several trials going on with different therapies to regrow the cartilage in a damaged joint, none of them are in market just yet. In the course of my study, I began reading about stem cell therapy and that encouraging results were being achieved in some people. Digging in further, I learned about the different types of stem cells, the different methods of obtaining them, and the various procedures.

I ultimately went with stem cells harvested from my iliac crest; that's the place in the body that retains the most active stem cells as you get older. Withdrawing the cells, spinning them down to condense them, and then injecting them into my hip joint. It's not the most pleasant procedure, but it beats surgery. I'm happy to say that a year-plus later, I'm walking again, skiing, and even playing pickleball. I won't say my leg is 100 percent, but it's probably at 85 to 90 percent. And that buys me time until the new cartilage therapies come into the market.

Beyond novel treatment to avoid surgery, stem cell therapy is disruptive to the medical industry. While stem cells are showing increased efficacy for a growing number of conditions, the U.S. Food and Drug Administration (FDA) has only approved the use of stem cells for a couple, very specific, conditions. But I don't think the problem is entirely with the FDA. The health insurance industry would sooner pay $35,000 for a hip replacement, rather than a tenth of that amount for stem cell therapy, to help protect healthcare revenue. And why not? When consumers are overwhelmed by the byzantine health industry, never sure who is paying for what, or what the true cost of procedures, therapies, and medications are.

It's that very complexity that has served to protect the system that creates an opening for a company like Amazon. The company's recent acquisition of One Medical gets Amazon into the healthcare business in a very direct and scalable way. Amazon Prime members can join One Medical for only $144 for the first year and receive on-demand virtual care, same-day or next-day office visits, walk in lab services, and more. The One Medical acquisition builds on Amazon's RxPass, a subscription providing access to more than fifty generic prescription drugs for a low five-dollar-per-month fee.

Amazon positions both the healthcare service and prescription drug service as being so inexpensive that customers do not need to bother submitting the costs to their traditional health insurance. Amazon's health-related innovation butterflies are certainly disruptive to the massive and complex healthcare industry.

I include this healthcare disruption example for two reasons. The first is that many grocery retailers have pharmacies within their stores and view the prescription drug business as a key part of their operation. The second reason is that many people within the fast-moving consumer goods retail industry feel that it is too complex to be easily disrupted. I think what Amazon and others are doing to disrupt healthcare can serve as a good example to retailers of the danger of complacency.

My experience, and Amazon's initiatives, are just a couple small examples of the disruption happening in healthcare. But, as we've discussed, the innovation flywheel is spinning ever-faster as we move up and out the exponential growth curve of technology, and there are more new "butterflies" emerging every day.

According to Statista, there are 1.35 million tech startups globally. And innovation is a global game, with fascinating technologies coming out of Israel, Australia, Germany, and many other markets around the world. While a lot of tech innovation continues to flow from Silicon Valley, its impressive to see innovative capabilities emerging from many areas across the United States, fueled by the work-from-home movement that developed during the pandemic, increased connectivity and network speeds, and a growing array of collaboration tools.

So, there's an interesting perspective here: Retailers are certainly at risk of being disrupted by new technologies flowing into the industry. But traditional retailers also have the opportunity to use new capabilities to become the disruptor, using new technologies—maybe like those discussed here—to disrupt their competitors.

Readers will have heard about some or even all of the following capabilities, a few of them for some time. And that's the problem. As we read about earlier, exponential growth is deceiving in the early stages—it's hard to detect, and many times it is ignored. But the technologies discussed here are either at, or will soon reach, the tipping point, the inflection point on the tech growth curve where explosive growth begins . . . and disruption follows. Pay attention to these and begin to consider how you can embrace these changes to avoid critical disruption of your business.

Interrupting Business as Usual

Think about the last time you went to a doctor. Chances are you had to provide your health history along with whatever medications you may be taking. If you're like most people, your health data is spread across multiple providers that you've visited over the years. Prescription data existing at different pharmacies, care data at the hospital you were in five years ago, prescribed physical therapy data somewhere else. Doctor records from years past in different cities where you may have lived. You get the idea because we've all lived it.

Having our health data spread all over is not only a headache each time we go to the doctor, but it can also actually be dangerous if pertinent health records can't be accessed in a timely way. Health care providers may make decisions on your treatment without seeing the full picture of your health history.

Now think about all the loyalty programs you belong to. Different retailers—grocery stores, drug stores, convenience stores, department stores, and more. Travel loyalty programs from the airlines, hotels, car rental agencies, and others. And then there are all the eCommerce sites you visit and interact with: Netflix and other streaming services, Amazon, and countless others. As you start adding all these up, you begin to realize just how many places are holding your personal identifiable information (PII).

Now think about how up to date that information is . . . or isn't. Do you trust all these different companies with your data? Does it make sense for a company you haven't done business with for some years to still have your information in their database?

And having your personal information held in so many different places increases the odds that a hack will result in your data being released across the dark web. That issue is driving the increased focus on privacy and is a growing factor in driving meaningful digital engagement.

A Better Way

Sir Tim Berners-Lee believes there's a better way. And he should know. Berners-Lee is the creator of the worldwide web (not to be confused with the internet), and he has grown increasingly concerned with how the web has developed with large companies owning, controlling, and monetizing individuals' data. That concern led Berners-Lee to co-found Inrupt, a company to realize his vision.

> *Berners-Lee's vision is becoming reality through the Solid project. Its core idea: Instead of your personal data being scattered all over the web, kept by whoever collected it from you, it should all be stored in one place and under your control. That place is called a Pod, which is short for "personal online*

data store." You can give any website access to information in a Pod—or, just as easily, revoke that access. Solid also offers a single sign-on feature, allowing you to log into any site that supports it, similar in concept to the universal log-in tools offered by Meta, Google, and Apple, without tying you to a tech giant's ecosystem.

Berners-Lee compares a new Pod to an empty Scrabble board. "But as your life progresses—your digital identity or digital state—apps write stuff in, and it gets more and more interesting, more and more rich," he explains. It also stays private, which he says makes it fundamentally different from Web3 and its use of public blockchain technology.

"Because Solid can store data of any sort, its potential applications are as sprawling as the web itself. In Belgium, for example, the government of Flanders has committed to using Pods as the basis of the region's entire digital economy. And in October, the BBC began using Pods to power its "watch party" feature. That lets viewers give the broadcaster access to additional data about themselves beyond what it knows from their streaming habits, allowing for more relevant content recommendations. As the BBC itself has pointed out, those suggestions could get even better if other streamers such as Netflix and Spotify were also to hook their algorithms up with Solid, allowing users to maintain records of their tastes that spanned multiple media platforms.[*]

---

* Harry McCracken, "Tim Berners-Lee Is Building the Web's 'Third Layer.' Don't Call it Web 3," Fast Company, November 8, 2022, *https://www.fastcompany.com/90807852/tim-berners-lee-inrupt-solid-pods.*

Now imagine what healthcare looks like in an Inrupt future. The individual having in one Pod all their healthcare history, from doctors' notes to prescribed therapies and medicines, along with up-to-date insurance information, even real-time health data flowing in from wearables. The individual then having the ability to share all or some portion of their data with the doctor they're presently seeing or the pharmacy where they're picking up a prescription.

Not only will health outcomes improve due to more complete health data but the power in the healthcare industry shifts to the consumer. The consumer in position to effectively say, "I'll give you access to my data and you have to provide me cost of care information in a fully transparent manner." Now that's disruptive.

Imagine what retail could look like. The shopper maintaining a single profile that they manage and control. That profile containing the usual name, address, email, etc. but also loyalty IDs, health-related information, and more. The shopper then having the ability to share some or all of their profile with a chosen merchant in return for some benefit.

Disruptive Data Ownership

What Berners-Lee is fixated on is the next evolution of the World Wide Web that he created more than thirty years ago. Web 1.0, also known as the "read-only web" or the "static web," refers to the early stages of the World Wide Web when it was primarily used for information retrieval and consumption. This era spanned from the inception of the web in the early 1990s until the late 1990s. During this period, websites were largely static and offered limited interaction or user-generated content. Web pages were predominantly text-based with simple graphics and lacked the dynamic elements we are accustomed to today.

In Web 1.0, the internet was mainly a one-way street where users passively consumed content. Websites were primarily designed for businesses and institutions to provide information, such as company profiles, product catalogs, and contact details. Interaction with websites was limited to clicking on hyperlinks to navigate between pages and read information.

Web 2.0 emerged in the early 2000s and marked a significant shift in the way people interacted with the web. This phase is often referred to as the "read-write web" or the "social web." Web 2.0 introduced a more interactive and collaborative online experience, enabling users to contribute, create, and share content with others.

Unlike Web 1.0, Web 2.0 fostered user-generated content and encouraged participation, leading to the rise of social media, blogging platforms, and online communities. Websites became more dynamic, incorporating multimedia elements, such as images, audio, and video. Users could now comment on articles, share their thoughts, and engage in discussions. Content creation and consumption became a two-way process.

Web 2.0 brought forth numerous platforms and services that allowed individuals to connect, collaborate, and share information across the internet. Social networking sites like Facebook, Twitter, and LinkedIn gained popularity, while platforms like YouTube, Flickr, and SoundCloud facilitated the sharing and discovery of multimedia content. E-commerce experienced a surge with the emergence of platforms like Amazon and eBay.

Web 3.0 envisions a paradigm shift in data ownership and control compared to its predecessors. In Web 1.0 and Web 2.0, user data was primarily controlled by centralized entities such as social media platforms, search engines, and online service providers. Companies like Google, Facebook (Meta), Twitter, and others, accumulated massive wealth as they aggregated, curated, controlled, and monetized all the data created by individuals. Web 3.0 seeks to empower individuals by returning data ownership, privacy, and control, to the person.

One of the key concepts associated with Web 3.0 is the idea of personal data sovereignty. It promotes the notion that individuals should have full ownership and control over their personal data. This means that users should have the ability to decide how their data is collected, used, and shared, as well as the ability to revoke or modify those permissions at any time.

Technologies such as decentralized identity (DID) and self-sovereign identity (SSI) play a vital role in Web 3.0's approach to data ownership. These technologies allow users to have unique, portable, and verifiable identities across different platforms without relying on centralized authorities. With SSI, individuals can manage and control their identity and associated data, thereby reducing reliance on centralized intermediaries.

Web 3.0 also embraces the concept of data interoperability, where different applications and services can securely and seamlessly exchange data with user consent. Interoperability ensures that users can freely move and share their data across various platforms, fostering competition, innovation, and user empowerment.

Inrupt is powering up this shift to Web 3.0, facilitating individual ownership of one's data. Think about what the world will look like in just a few years as Berners-Lee's vision becomes a reality. Every consumer owning and controlling their information and content.

Think about all the retailers with loyalty programs. And how many of those retailers are leveraging and monetizing their shoppers' data. Not all that different from how Meta (Facebook) or Google monetize users' data.

That shift in the control and ownership of consumer data is positioned to blow up the hottest trend in retail today: retail media networks. Brand advertisers are projected to spend over $60 billion in 2024 in retail media networks, with a vast majority of that money flowing into the networks of the largest retailers, companies like Amazon, Walmart, Kroger, Target, and a handful of others.

What attracts advertisers to retail media networks is the ability to leverage massive data to create and target audiences, oftentimes using shopper identified transaction data. The retailers execute campaigns using shoppers' personally identifiable information (PII) data, including their email address or home address. And then close the loop with full attribution, enabling the advertiser to connect the dots from audience to ad targeting to ad views to product purchase. But all of that goes away in the world of Inrupt unless the shopper gives permission to the retailer or the brand advertiser to use their data. And do you think shoppers may want something more than a fifty-cent coupon in return for the use of their data? It's going to be a whole new world.

Inrupt represents one of, if not the, most disruptive of the emerging butterflies. I have believed for a long time that there must be an ultimate shift to the consumer/shopper, giving the individual control of their data and information. Inrupt's technology appears to be the path to making that happen, with forward-thinking retailers becoming a trusted partner into the future.

Imagine, for example, if you're a food retailer helping your shoppers use Inrupt's technology to consolidate all their food purchase and their health data. The retailer, as a trusted partner, can ask the shopper to share (limited) access to their health data so as to provide guidance to beneficial and relevant foods.

I think the opportunity for retailers here is to be seen as advocating for the customer and bringing capability to the customer to help them regain control of their data. Beyond that is the opportunity to develop new services and partnerships in the world of Web 3.0.

The Rise of AI

> *"Business models are about to change across all industries, and the technology that is driving this is the widespread use of artificial intelligence everywhere."*
>
> *—Peter Diamandis*

AI is being woven into nearly every capability, making solutions smarter and more effective. It is a foundational technology in the sense that it underpins a growing number of exponentially growing capabilities like voice, computer vision, robotics, process automation, and more.

And AI is growing fast. According to a study done by Stanford University, computing power for artificial intelligence is now doubling every 3.4 months. That is exponential growth . . . and fifty to sixty times faster than Moore's law. Here's an example:

"A classic test of artificial intelligence is determining whether a photo was of a dog or of a cat. In the beginning of 2017, it cost $10,000 to classify one billion images on the latest computers. By the end of 2019 it cost 3 cents to perform the same task."*

A December 2019 Stanford report, produced in partnership with McKinsey & Company, Google, PwC, OpenAI, Genpact, and AI21Labs, found that AI computational power is accelerating faster than traditional processor development. "Prior to 2012, AI results closely tracked Moore's law, with compute doubling every two years.," the report said. "Post-2012, compute has been doubling every 3.4 months."†

All that computing power is making AI smarter . . . fast.

> *"These AI systems are now capable of passing many standardized tests, from high school to graduate- and professional-level exams that span mathematics, science, coding, history, law, and literature. Google's Med-PaLM performed at an 'expert' doctor level on the medical licensing exam, not only correctly answering the questions but also providing a rationale for its responses. The rate of improvement with these systems is astonishing. For example, GPT-4 made significant progress in just*

---

* Matthew Boutte, "The Future Is Going to Happen a Lot Faster than the Past Did," Medium, August 24, 2020, https://medium.datadriveninvestor.com/the-future-is-going-to-happen-a-lot-faster-than-the-past-did-8bc7295f63ef.

† Cliff Saran, "Stanford University Finds that AI is Outpacing Moore's Law," Computer Weekly, December 12, 2019, https://www.computerweekly.com/news/252475371/Stanford-University-finds-that-AI-is-outpacing-Moores-Law.

four months, going from a failing grade on the bar exam to scoring in the 90th percentile. It scored in the 93rd percentile on the SAT reading and writing test and the 88th on the LSAT and got a 5—the top score—on several Advanced Placement (AP) exams.

"AI models like GPT-4, Microsoft's Bing Chat, and Google's Bard are advancing beyond simple knowledge repositories. They are developing into sophisticated reasoning engines that can contextualize, infer, and deduce information in a manner strikingly similar to human thought. While traditional search engines functioned like librarians guiding users toward relevant resources, this new generation of AI tools acts as skilled graduate research assistants. They can be tasked with requests such as conducting literature reviews, analyzing data or text, synthesizing findings and generating content, stories and tailored lesson plans.

Artificial intelligence is advancing quickly. AI systems are already outperforming human beings in a growing number of areas like language understanding, reading comprehension, and image recognition. These three areas help AI learn and so are serving to further accelerate AI capabilities.

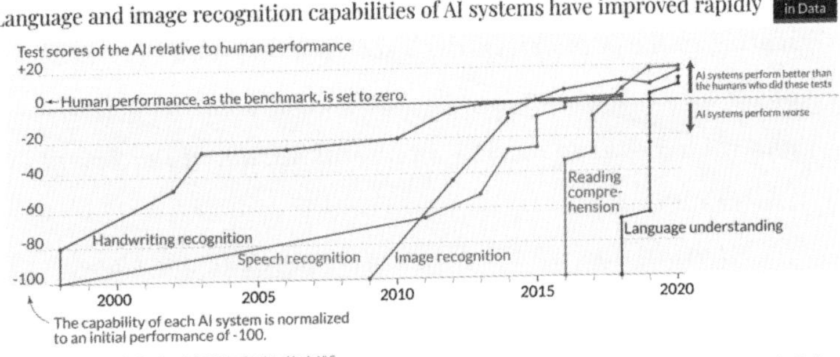

Peter Diamandis, executive founder of Singularity University, suggests AI capability will surge. "As predicted by technologist and futurist Ray Kurzweil, artificial intelligence will reach human-level performance this decade (by 2030). Through the 2020s, AI algorithms and machine learning tools will be increasingly made open source, available on the cloud, allowing any individual with an internet connection to supplement their cognitive ability, augment their problem-solving capacity, and build new ventures at a fraction of the current cost."*

Voice

Introduced in the 2008 movie *Iron Man*, Jarvis is a voice-activated computer program embedded in Tony Stark's high-tech home. Jarvis, an acronym for "Just A Rather Very Intelligent System," evolves through the movie series to become a sophisticated, multi-functional

---

* Peter H. Diamandis, "Metatrends Shaping the Next Decade," Diamandis.com. November 12, 2020, https://www.diamandis.com/blog/metatrends-shaping-decade.

AI and an integral part of Stark's technology arsenal. At the rate artificial intelligence is growing, we'll each have our own Jarvis soon.

"As services like Alexa, Google Home, and Apple Homepod increase their capabilities, they will expand to become part of our lives 24/7, serving as our interface with the world around us. Imagine a JARVIS-like 'software shell' that you give permission to listen to all of your conversations, read your email, and monitor your blood chemistry.

"With access to such data, these AI-enabled-software shells will learn your preferences, anticipate your needs and behavior, shop for you, monitor your health and help solve your problems in support of your goals. Your JARVIS-like software will call an autonomous uber for you whenever needed, before you have to ask, will alter the room's temperature, and serve up your favorite food before you ever ask. Imagine the world's best 'executive assistant' who is all-knowing and almost psychic as it meets all of your needs and desires."*

We're already experiencing some of this future as voice technology becomes more pervasive. We have smart assistants on our smartphones, on our smart watches, in our cars, and a rapidly growing array of devices in our homes, from smart speakers to televisions to home security systems.

Voice-enabled services are proliferating everywhere. Healthcare is using a growing array of voice-based capabilities, from smart speakers in hospital rooms, to doctors offices for recording notes automatically. And voice is expanding into the office. Just this week

---

* Singularity University, "Abundance 360: Top 20 Metatrends & Moonshots for 2022-2032," accessed September 2023, https://www.diamandis.com/hubfs/METATRENDS%20%26%20MOONSHOTS%20-%20V2.pdf?.

a young company approached me with a voice-to-SQL query capability, letting users simply ask for ad hoc analysis against massive data sets—no skills required. Voice driven picking has long been used in distribution centers and has moved into the store to support picking online orders. McDonald's, White Castle, Dunkin Donuts, and Wendy's are all testing AI powered chatbots in their drive-throughs.

"On the heels of Panera's debut of Amazon's contactless palm payment solution, the fast-casual bakery café chain is also partnering with Amazon in other ways. MyPanera members will now be able to order via Alexa-enabled and Echo Show devices, adding yet another perk to the Panera rewards program.

Amazon device users can simply ask Alexa to add, customize, and order Panera items for pickup or delivery, using a stored payment method and delivery address in their MyPanera account. Panera is the first brand to use Alexa's updated food skills API, which introduces conversational AI, making it less frustrating to interact with and successfully order food via a robot voice. The technology understands order modifications like 'extra bacon' or 'hold the avocado,' which makes it part of the trend of upgraded voice AI that understands colloquialisms, changes in speech patterns, and mid-conversation pivots."* Amazon has even taken Alexa shopping, installing Alexa stations at various locations in its Amazon Fresh stores. Alexa can help shoppers find products, suggest recipes, and more.

Blutag is a young company bringing voice capability to e-commerce systems, letting shoppers add items to their list, find recipes, and more, all by speaking. The Open Voice Network is a non-profit organization developing technical standards and guidelines for the future of conversational AI. Their work highlights the growing role of AI-powered voice capabilities showing up in all parts of our lives.

Health & Wellness

Consider this: An estimated 60 percent of adults in the United States have at least one chronic health condition. Forty percent of adults have two or more chronic diseases.† This, while the rate of obesity in

---

* Joanna Fantozzi, "Tech Tracker: Omnichannel Accessibility Keeps Expanding in News Ways," Nation's Restaurant News, April 17, 2023, https://www.nrn.com/technology/tech-tracker-omnichannel-accessibility-keeps-expanding-new-ways?NL=NRN-014&Issue=NRN-014_20230427_NRN-014_575&sfvc4enews=42&cl=article_1&utm_rid=CPG06000000263787&_mc=em_NRN_News_NRN%20Technology%20Report_News_NL_04272023_49211&utm_campaign=55208&utm_medium=email&elq2=8e3977cf9e504cb9b89b28a41616ae4a&sp_eh=e67ed679c944c6aeecc5d56dcdafbc8b77e72a779e514a0169520e2592b84d98.

† Erin L. Boyle, "The Cost of Being Chronic in 2023: A Special Report," HealthCentral, April 28, 2023, https://www.healthcentral.com/chronic-health/the-cost-of-being-chronic-in-2023-a-special-report.

the United States is nearly double that of other developed countries.*

Coming from a career spent in the food industry, the link between health and the foods we consume is something of great interest to me. It is shocking that life expectancy in the United States has declined for two years in a row—this is happening in the world's richest country. Given the United States spends more on healthcare than any other country, and yet lags in life expectancy, should give us all pause.

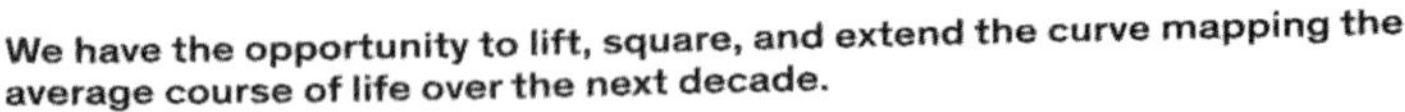

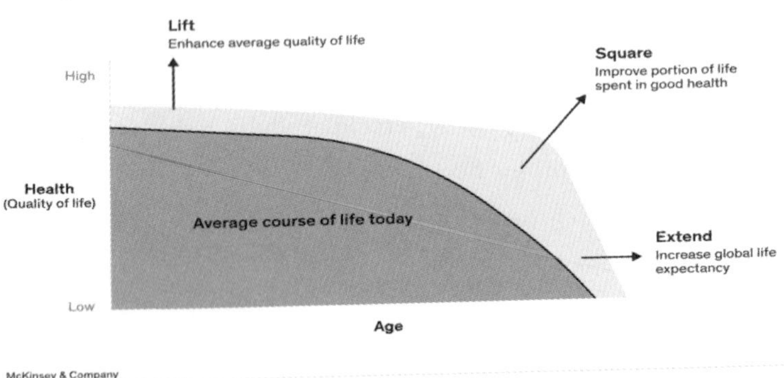

And the problem is worse than it appears on the surface. A McKinsey report calls out that, "on average, people spend about 50 percent of their lives in less-than-good health including 12 percent

* The Commonwealth Fund, "U.S. Health Care from a Global Perspective, 2022: Accelerating Spending, Worsening Outcomes," January 31, 2023, https://www.commonwealthfund.org/publications/issue-briefs/2023/jan/us-health-care-global-perspective-2022.

of their lives in poor health."* The report goes on to say that the problem appears to be getting worse, especially in high-income countries like the United States, where chronic health conditions are becoming more prevalent for a longer time. Indeed, diet is the number one cause of poor health and death.

The U.S. government funds healthcare for a growing portion of the population to the tune of billions of dollars each year. And at the same time, the government guides consumers to foods that are not healthy through the suggested dietary guidelines. As healthcare costs grow to a projected 20 percent of the U.S. economy by around 2030—over $6 trillion dollars—something has to give.†

---

* Erica Coe, Martin Dewhurst, Lars Hartenstein, Anna Hextall, and Tom Latkovic, "Adding Years to Life and Life to Years," McKinsey & Company, March 29, 2022, https://www.mckinsey.com/mhi/our-insights/adding-years-to-life-and-life-to-years.

† https://www.healthaffairs.org/doi/10.1377/hlthaff.2023.00403#:~:text=National health expenditures are projected to have grown 4.3 percent,to 17.4 percent in 2022.

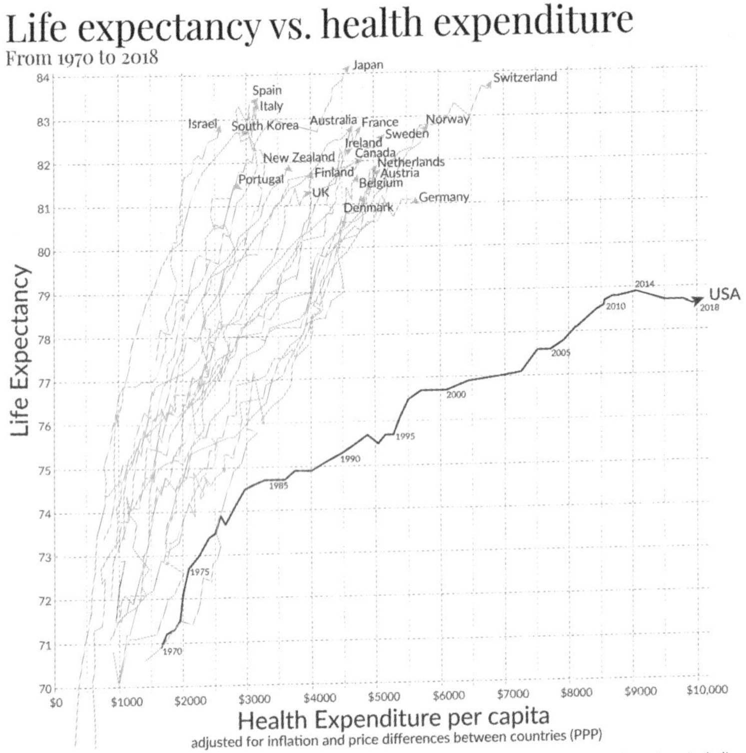

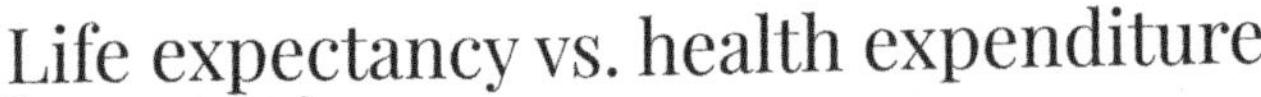

Data source: OECD – Note: Health spending measures the consumption of health care goods and services, including personal health care (curative care, rehabilitative care, long-term care, ancillary services, and medical goods) and collective services (prevention and public health services as well as health administration), but excluding spending on investments. Shown is total health expenditure (financed by public and private sources). Licensed under CC-BY by the author Max Roser. OurWorldinData.org - Research and data to make progress against the world's largest problems.

And that's where two innovation butterflies are emerging, one more near term, the other a little ways off, but both connected.

Personalized Nutrition Guidance

The first butterfly is personalizing nutritional guidance for an individual based upon their specific health conditions and concerns; this technology is able to help a shopper with diabetes or a different shopper with high blood pressure understand what specific foods they should be purchasing to improve their well-being. Think of it like having a personal dietitian accompany you on your shopping trip through the store.

This kind of personalized guidance to relevant foods presents a significant opportunity for retailers to build trusted relationships with their customers and be seen as a valued partner. This space is moving quickly with more capabilities coming into the market around recipes, meal planning, food product attributes, and guidance. And this provides a natural segue for the retailer to build partnerships into healthcare in their marketplaces, helping bring all the pieces together on behalf of their customers.

Sifter Solutions is leading the way, powering sophisticated, AI-driven food guidance to shoppers. Using their technology, a shopper begins by creating their personal profile, using hundreds of dietary filters, ingredient exclusions, and lifestyle preferences. Sifter then analyzes every grocery food product against the personal profile created. And as we know, these profiles can be incredibly complex in today's world. The application makes it easy for shoppers to create their shopping list by diet, discover relevant recipes, and more, both in the store and when shopping online.

As called out in several places through this book, data—good data—is foundational to providing this kind of capability. Sifter ingests product ingredient data from multiple sources, cleanses the data, and then organizes it, using their AI capabilities to tag each food product with the different attributes that link back to the shopper's profile.

The Sifter team leverages nutrition standards from leading health organizations and has a comprehensive quality assurance methodology resulting in a complete and accurate food nutrition database. Sifter powered up Walmart's 2022 Wellness Day event and is being integrated to KVAT Food City's website and eCommerce, along with many other retailers.

And here's where it gets interesting: Sifter powers up SNAP-eligible food products for a growing number of food-as-a-benefit programs and is being woven into a growing number of tools offered by health insurance companies like Ayble Health, helping their subscribers easily discover foods beneficial to the individual's health concerns.

ScriptSave has brought to market their Personalized Wellness product that scores every food product across a participating supermarket to a growing number of health conditions. While the data science behind this certainly qualifies as an innovation butterfly, it is where this space is going that makes it transformative.

ScriptSave has entered into a significant pilot with Elevance Health (formerly Anthem Insurance) to make the Personalized Wellness capability available to diabetic subscribers who are on Medicare. And in one of the pilot markets, Louisiana, the regional grocery retailer Rouses Markets, is joining in. The ultimate goal is to bring participating subscribers additional savings on food products that will make them healthier.

ScriptSave's capability brings together detailed product ingredients with cutting edge nutrition science to power those food scores. It is understood Kroger is working on such a project internally, expanding on the OptUP app that today provides general nutrition scores.

I believe we will see an explosion of activity around this space over the next couple years as Kroger, Walmart, and other retailers, partner with health insurance companies, big payers (large employers who self-fund healthcare for their employees), and health care providers.

In a sign of the growing food–health convergence, Meijer has created a partnership with Blue Cross Blue Shield of Michigan to create a new Medicare plan. The plan includes a $0 monthly premium and a $660 annual allowance to spend across the 120 or so Meijer stores in the Michigan market. Shoppers can use the allowance to purchase healthy foods if they have a chronic medical condition.*

Human Longevity

The next emerging butterfly is around human longevity. There are a growing number of experts who believe that productive human lifespan can be extended to 100, 120, even 150 years. The goal is not to just extend lifespan, but to make those years healthy and productive for the individual. There are countless advances being made almost daily toward that end in all industries and spaces, not just retail.

"A dozen game-changing biotech and pharmaceutical solutions (currently in Phase 1, 2, or 3 clinical trials) will reach consumers this decade, adding an additional decade to the human health span. Technologies include stem cell supply restoration, wnt pathway manipulation, Senolytic Medicines, a new generation of Endo-Vaccines, GDF-11, supplementation of NMD/NAD+, among several others. And as machine learning continues to mature, AI is set to unleash countless

---

* Alarice Rajagopal, "Meijer Gets into Health Insurance with New Blue Cross Partnership," Supermarket News, October 2, 2023, https://www.supermarketnews.com/nonfood-pharmacy/meijer-gets-health-insurance-new-blue-cross-partnership.

new drug candidates, ready for clinical trials. This Metatrend is driven by the convergence of genome sequencing, CRISPR technologies, AI, quantum computing, and cellular medicine."

There is growing study and understanding of the role of food—or, the lack of it—in human longevity. Increasing numbers of people are using intermittent fasting to reduce the calories they consume, as different studies establish a link between calorie reduction and lifespan.

Longer Life Implications

As we live longer, more productive, and active, lives, retailers need to consider the implications. Are containers easy to open? Is signage text large enough to read easily? Is food guidance easy to access? Are the products and services provided aligned to an older population? Will 100-year-old shoppers want to navigate a 100,000-square-foot store?

Intermittent fasting is growing in interest as a way to both control weight and extend lifespan. Another path are weight loss drugs like Ozempic, Mounjaro, and others coming into the market that help suppress a person's appetite. A recent *Wall Street Journal* article projects "nearly 7 percent of the population may be on weight-loss drugs by 2035."* And already big food manufacturers are watching if suppressing appetites will be reflected in less food purchases.

And there are implications for retailers as employers. How can retailers take advantage of employee experience ten years, or even twenty years, past traditional retirement age? Is there an impact to

* Jesse Newman, "America's Food Giants Confront the Ozempic Era," *Wall Street Journal*, October 5, 2023, https://www.wsj.com/business/ozempic-impact-snack-food-companies-9eec87e5?mod=hp_listb_pos2.

retirement plans and finances? How is rapidly changing technology handled by older employees? Do more physical limitations need to be taken into account?

There are countless implications to society and businesses as the vision of longer, more productive, lives is realized. And it's not too soon to be thinking about it. Healthier, more productive, and longer lives. What could be a more beautiful innovation butterfly?

Chapter Summary

- These are the massively disruptive butterflies emerging that promise to dramatically alter how we engage and interact with the world around us.
- Although retailers need to be constantly vigilant regarding disruption caused by competition, there are also key opportunities for you to become the disruptor in your industry.
- Some innovation butterflies provide disruption with many opportunities for positive benefits if organizations ride the waves of change, these include expectations from consumers for personalized nutrition guidance as well as adapting store experiences to accommodate people with extended life spans.

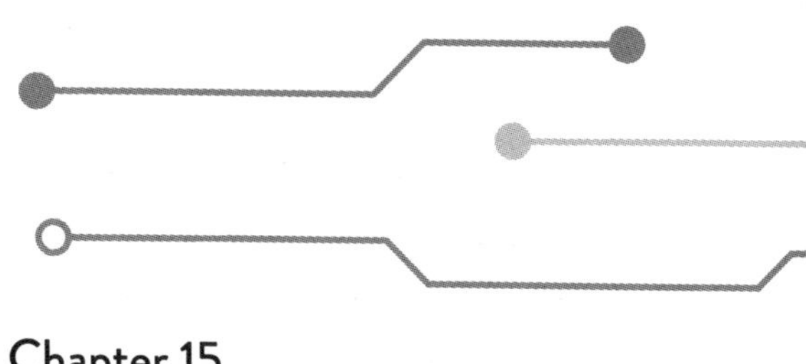

Chapter 15

THE PATH FORWARD

Alice never could quite make out, in thinking it over afterwards, how it was that they began: all she remembers is, that they were running hand in hand, and the Queen went so fast that it was all she could do to keep up with her: and still the Queen kept crying "Faster! Faster!" but Alice felt she could not go faster, though she had not breath left to say so. The most curious part of the thing was, that the trees and the other things round them never changed their places at all: however fast they went, they never seemed to pass anything. "I wonder if all the things move along with us?" thought poor puzzled Alice. And the Queen seemed to guess her thoughts, for she cried, "Faster! Don't try to talk! If you want to get somewhere else, you must run at least twice as fast as that!"

—*Lewis Carroll,* Through the Looking Glass

Running faster and faster to simply stay in the same place seems an apt analogy for the retail industry as innovation accelerates. Every retailer whom I speak with is trying to do more with the same, or sometimes less, resources. Staffing shortages, growing demands for technology investment straining financial resources, everything at a faster and faster pace. Compounding the issues are the growing misalignment between traditional retail organizations and business models and organizations and models built to thrive in an exponential world. As was called out earlier, retail's wheels are starting to come off.

Retail's response to ever-faster, ever-broader tech-fueled innovation has been to try and do more, faster. There comes a point where demands simply outpace existing capability. While legacy systems, processes, and organizational structure, all constrain retailers from freely innovating, at the heart of it is culture. To move beyond those constraints, retail must look to more fundamental change if they want to ensure not only sustained success but survival in this new era.

Much of this work is centered around the three steps that have been called out in several places throughout the book.

1. **Developing an inspired vision.** Identify and set aside the personal and organizational beliefs that impede creating an inspired vision for the future of retail.
2. **Committed action.** Take purposeful, deliberate, action to move your company to the exponential path to the future. Learn about and implement tech-first and first-principles thinking.
3. **Creating—and living by—an innovation doctrine.** Your innovation doctrine should document the principles, strategies, and guidelines that help govern the organization. And then embed the doctrine in your organization and culture or risk your company's "immune system" rejecting new innovation and new practices.

Organization Transformation

Becoming a retail organization able to thrive in the exponential world requires work. And for retailers of any size, thriving requires intentional, committed, action from the board of directors, the company's C-suite and executive team, middle management, and front-line workers.

Board of Directors

Directors have a fiduciary duty to the company and its shareholders. As such, they need to understand the new world that retail has entered past the inflection point. Entering the exponential world brings with it increased risk, along with vast opportunity. But addressing both those things requires change and investment. Alongside that understanding is creating an imperative to change and an urgency to transform the company. Ideally, work and education with the board helps gain and build support for the CEO and executive team.

C-Suite and Executive Team

Beyond the understanding and imperative to change, work at the executive level needs to focus on transforming the company's culture—no easy thing. This includes changing beliefs, consideration of new possibilities, inspired vision, and first-principles thinking. We think it's important to create an innovation doctrine for the organization.

Beyond this company transformation, it's important to work on making resources available to discover and learn about new innovation along with creating an incubator or using stores as laboratories to test and trial new capabilities.

Key Executives / Junior Executives / Rising Stars

Obtaining buy-in from this group is critical to changing paths, moving from the linear path of today to the exponential path of tomorrow. Gaining understanding of how fundamentally retail is changing helps establish the foundation for discussion around the transformation that must take place: changing beliefs, welcoming new possibilities, and aligning vision with committed action.

The big goal is to proactively suppress the immune response from the organization to the fundamental change required. We would also suggest drawing talent from this group to set up and operate incubators or stores-as-labs.

Everyone Else

The most effective leaders understand the importance of bringing everyone along with them as they embark on a new course. Changing paths to setting a retail organization on a course to the exponential future can be threatening to many people with jobs changing, long-established processes taken over by AI apps, and more. Planning needs to happen around how you're going to educate and bring your associates from across the organization along with you into the future.

It's important to realize that the future can be scary. Workers across industries are increasingly worried about the potential threat of new technology to their jobs. Physical automation threatens physical work. AI-powered business process automation threatens workers in merchandising, marketing, human relations, and more. Communicating an understanding of this concern is important. Better is making workers across the enterprise partners in the journey to creating the future of retail.

Innovation as a Process

New, transformative technology is no longer something that comes along every few years. There are new butterflies coming into the market almost daily, powered by the innovation flywheel. Technology-fueled innovation must be viewed as an ongoing process, which is something lacking at most retailers I encounter. I typically see several issues that retailers suffer from:

1. Lack of a comprehensive technology innovation plan. Too many retailers take a scatter-shot approach to acquiring and deploying new capabilities, driven by what their competitors are doing or whatever tech has the latest 'buzz'.
2. Some retailers have a list of capabilities they want to secure but go about it like checking off the next thing on their to-do list. Often, this relates back to the first point around planning.
3. Many retailers rely on existing vendors to bring them new innovation and fail to understand what else may be out in the market. With new solutions being made available all the time, retailers would be well served to do some research before signing an agreement.

An innovation process needs to incorporate several elements:

- **Awareness and Discovery:** Traditional retailers of any size are simply not built for discovering the countless new capabilities flooding into the retail industry. So many executives are simply overwhelmed by tech companies reaching out for an audience, email boxes, voicemail systems, maxed out.
- **Evaluation:** Which of the new capabilities you've become aware of are relevant to your organization and your plan to

create the future? Are you considering distribution center automation? Or computer vision platforms? Or marketing personalization capabilities? With more new innovation flooding into the industry, we would suggest developing some criteria for evaluating which solutions you want to progress with vs. those not aligned. Things to consider are cost (both initial and ongoing), projected ROI, references from other retailers that use the solution, integrations with other systems, and so on.

- **Due Diligence:** Develop a scorecard for due diligence on companies you want to potentially pilot. Things like the experience of their team, financial resources, other deployments in retail, the robustness of their technology, and alignment with your company's culture and values, should be considered.
- **Pilot:** Of the companies that pass due diligence, which of them should be selected to pilot? This also involves building out all the details of a pilot, where, when, how, who is funding the pilot, and more. Most importantly who is responsible for overseeing the pilot and what are the success metrics or criteria?
- **Deploy:** What does deployment look like for capabilities that make it through the pilot stage? What will it take to deploy a given capability? Does it support internal operations? What training is required, etc. Or is it shopper facing and require marketing involvement to communicate?

A New Approach to Innovation

FMI is the Food Industry Association, and it represents the country's supermarket retailers and a growing number of brand manufacturers. At one time, FMI organized a massive annual industry convention, the FMI Show, attended by tens of thousands of industry people from across the United States and from around the world. In the late 1990s, as technology was becoming more important to retail, FMI launched Marketechnics, an annual conference dedicated to technology in retail, with a heavy focus on marketing tech.

Transformation and disruption also impact conferences and conventions, like FMI's. They made the decision some years ago to discontinue the large annual gatherings as industry consolidation and budget constraints began to impact attendance. But FMI took advantage of how things are changing to radically change their approach to the traditional trade show. FMI partnered with my organization, CART, the Center for Advancing Retail & Technology, to re-envision Marketechnics, with a goal of helping their members keep pace with accelerating innovation.

The new Marketechnics program was designed to also capitalize on another important trend: **relevancy**. The program is designed for the individual retailer (or wholesaler or brand manufacturer) who is the host. Think of it as a personalized trade show, featuring relevant innovation, just for the specific host (retailer). Once a host has committed to the program, the team spends time with the retailer's executive team to understand areas of technology interest, opportunity, or challenge. With those areas identified, the team then starts bringing together potential tech providers to participate, evaluating each potential solution provider against the retailer's goals.

Once the roster of participating tech companies is finalized with the retailer—typically, it is twelve to eighteen companies—the team then focuses on preparation, spending time with each to review their presentations to ensure alignment with the retailer.

The culmination of the program comes in a one- to two-day program on site at the retailer's headquarters. A typical agenda looks something like this:

Day 1: An industry briefing on the latest tech trends across the retail delivered to the retailer's executive team. Time allowed for questions and discussion. Also on Day 1, a mini-trade show is organized with each of the participating solution providers setting up a small display. The retailer invites appropriate personnel to spend time going through to talk with and learn about the new innovative capabilities.

Day 2: This is devoted to getting each of the tech companies in front of the retailer's executive team. The retailer gathers all their exec team and any other associates to hear short presentations from each of the chosen tech companies with time allowed for questions and discussion. There is a lot of networking that happens during coffee breaks.

The power of this approach to innovation lies in several things:

The first is aligning the tech companies chosen to participate with the retailer's needs, making each presentation highly relevant. Retail executives are busy people; this program distills down the best capabilities to present.

The second big impact is the trade show, especially for those retailers that open the show up to their associates. One retailer invited their entire tech team, along with other executives from across the company, to attend the trade show, which was organized around a happy hour. This kind of approach helps the retailer's team understand new technology and helps bring them on board with new ideas and approaches.

The real secret-sauce is bringing the retailer's entire executive team together for a day focused on learning about, discussing, and ultimately making decisions around new innovation. Everyone hearing the same presentation at the same time is invaluable. The ability of the CFO to weigh in with questions around some new marketing capability, or the COO asking questions about deployment of some new technology, is powerful. This approach dramatically reduces time-to-decision around new innovation and speeds up the entire process dramatically.

Our belief is that, at the very least, retailers should host an innovation program like this annually; the largest retailers could host several programs a year, each program focused on different areas like supply chain innovation, shopper engagement innovation, and so on.

Innovation in Action

Zach Bello has spent a good part of his career as part of, and leading, retailer innovation teams. He was involved with many of Kroger's innovation initiatives and pilots before moving to Hy-Vee where he led digital strategy and innovation. Zach helped coordinate two innovation programs like those discussed in this section and provides some valuable insights from his unique experience. He says:

Everybody's Job

Leading an innovation team means being at the forefront of a lot of really cool and exciting technology (like Nuro and autonomous grocery delivery). However, that doesn't mean that the innovation team is the only team responsible for innovating. The role of the innovation team isn't to be the only team innovating, it's to help provide structure and process to evaluating, measuring, and implementing solutions that help solve the problems facing the organization's biggest issues. In our early days really creating structure around an "Innovation Team," many teams would become upset, because that meant they were no longer able to "innovate," so making it clear that this team exists to help foster and drive innovation throughout the organization is key to ensuring everyone is involved. At the end of the day, those closest to the problems will be the most valuable, so it takes a collaborative effort across all teams to run a successful innovation program at scale.

Define Your "Why"

Aligning your efforts to actual opportunities will not only help gain support from the teams you will ultimately depend on to bring solutions to life, but also help you know whether or not things actually worked. This requires a whole other level of vulnerability that most organizations aren't accustomed to, as most organization don't reward people for being wrong (which, if you are running an Innovation program well, you'll be wrong more often than not). Having a well-thought-out hypothesis is key to knowing whether or not the things you are investing your time and resources in to was actually worthwhile.

Embrace Being Wrong

This goes across all levels of the organization, and creating and fostering a culture by which people don't feel like they've failed because they were wrong about something. While most Innovation teams operate in a world of mostly unknowns, it's impossible to get it right 100 percent of the time (or even the majority of the time). If your teams aren't comfortable being wrong, you run the risk of teams not setting good enough goals (e.g., goals they know they will accomplish), or won't take the risk entirely (not to be mistaken for carelessness). Being wrong means you've learned something new, which in turn will help you ultimately find the right solution for the problem you are trying to solve.

Measure Measure Measure

Sure, lots of things just seem like "cool" things to do, or follow a major trend, but always make sure you are measuring towards something. Define up front how you will determine how to proceed after you've implemented. We followed a "Pivot, Pause, or Scale" approach when running projects, based on how we were tracking against our hypothesis. We'd either continue piloting and pivot on something that we had learned to see if that changed the outcome, pause because we had everything we needed to know not to move forward, or needed time to analyze, or scale to bring the solution to life. This structure makes it easy to understand where your projects sit, how and why they got there, and also gives you something to refer back to when the project inevitably comes back several years down the road.

Zach's insights reinforce many of the messages communicated in Part 2 and are all critically important ingredients in retailers changing paths to successfully participate in the exponential future.

Innovation Implications

Part of preparing for a future arriving sooner than expected is to think about the implications of new technologies and capabilities.

Consider this scenario: You're getting out of your car in the parking lot of your favorite retailer and you're wearing your smart glasses. As you look at the store entrance, the retailer welcomes you to the store by name. But glancing up, there, over the store, is a virtual message from a competing retailer, inviting you to visit their store down the street.

The question is this: In a world of smart glasses and augmented reality capabilities, who owns the digital property rights in the air over the store? Does the retailer, who owns the store, also own the air rights over the store? If so, to what altitude? Does someone else own that digital "property"? Who?

As you can begin to appreciate from that simple example, the world of creating the future brings with it complexities that were never before thought about. Think about what Inrupt is doing, shifting control of an individual's data to the person. Who owns the transaction data identified to the shopper in the store or online? Who owns the patient's medical data from their last doctor visit? What privacy issues are raised by this kind of shift? Liability issues?

Or think about voice technology and the implications of Siri or Alexa overhearing—and recording—conversations going on in your home? If the virtual assistant overhears a crime being planned, is there a legal obligation to report it to authorities? You can imagine countless other scenarios.

Beyond more exotic legal implications there are more mundane considerations. A growing number of retailers are extending personalized savings to their shoppers. This implies an ability to deliver shopper-specific discounts in the checkout process. The existing digital coupon solutions or loyalty solutions are only capable of managing hundreds or maybe a few thousand digital offers. In a retailer with 50,000 product SKUs and millions of shoppers, the number of personalized promotions that must be managed explodes exponentially. Yet few if any retailers have this capability.

Time and again, I have seen retailers tie themselves in knots by entering into agreements with different solution providers that restrict the retailer's ability to do something else in a related area. Many times, this is not rocket science—it's a lack of attention to considering the implications of, often new, capabilities.

Chapter Summary

- Becoming a retail organization able to thrive in the exponential world requires work. And for retailers of any size, thriving requires intentional, committed, action from the board of directors, the company's C-suite and executive team, middle management, and front-line workers.
- Technology-fueled innovation must be viewed as an ongoing process that incorporates several elements, including: awareness and discovery, evaluation, due diligence, pilot, and deploy.
- Developing and supporting innovation programs within your organization is essential, but there are several considerations to keep in mind: (1) innovation should be everyone's job within an organization, not just the innovation program team's; (2) define the "why" behind your efforts to direct your

attention to impactful activities; (3) embrace being wrong because it teaches you something new to move forward with; and (4) measure progress and impact to provide a reference for future decision-making.

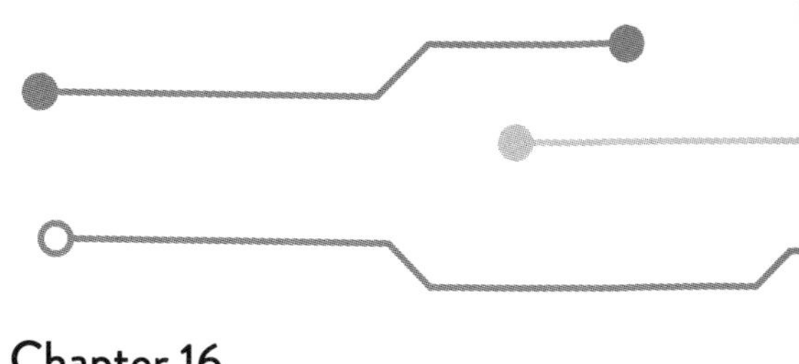

Chapter 16

CONCLUSION

My family's store started as a summer farmstand by my great-grandmother in 1934. I can remember as a kid getting up early on Saturday mornings to go to the regional market with my father, looking to buy fresh fruits and vegetables for sale that day, supplementing what we grew on our farm. My favorite part was swinging by a local bakery to pick up donuts and baked goods to sell. There's nothing like a donut still warm from the fryer, just dripping with glaze.

What I also remember is how my grandfather seemed to know everyone. He knew Alice, and that she shopped on Wednesdays and loved fresh, local strawberries still warm from the field. He knew Ben, and that he liked his strip steaks cut extra thick. And even Mrs. Johnson and her dog Buck, a massive Great Dane, who loved getting a bone as a special treat from our meat department. I remember we shopped for Mrs. Gardner, delivering her groceries on our way home. And this was years before home delivery was a thing!

That was personalized retail. The customer was the most important part of the business.

And that was the constant refrain over the ensuing years. "The customer is the most important part of our business" is the pablum I, and many others in the industry, was raised on.

Except something happened between then and now. Stores became bigger and busier. Product assortment exploded. Competition developed from every quarter. And food became available anywhere and everywhere.

As the industry grew larger, we began to lose focus on the customer. How we made money—or at least how we thought we made money—began to shift. Merchandising—deciding what products to put on the shelves and which products to promote—became an exalted position in our organizations.

The increasing focus on products grew hand in hand with a growing focus on obtaining marketing funds from brand manufacturers. One could easily make the case that it was because of marketing funds that products grew to take center stage. Today, many major retailers would not be profitable were it not for those subsidies that support sale prices to the shopper, help offset advertising costs, bolster category margins, and enrich the bottom line.

In a very real sense, brand marketing funds have insidiously shifted retailer focus away from the customer to a dependency on products. Retail became product driven.

But it doesn't have to be that way.

We as retailers can reclaim our customer heritage. We can use new technologies and capabilities to recreate those personal relationships of yesterday and become truly relevant to each of our individual customers today.

And that's what Retail in the Age of I is all about. It is about fulfilling the destiny of retail by regaining a focus on the customer. It is about building relationships with each and every one of our individual customers. And it is about returning products to their rightful role in service to each of our customers.

—*Retail in the Age of I*

That's an excerpt from my last book, *Retail in the Age of I.* The book was about how the world around us is becoming increasingly tailored to each of us individually. We have long taken for granted the personalization and relevancy provided us in the digital realm. Technology is increasingly enabling the physical world around us to be customized to our wants and needs.

3D printing is expanding quickly, clothing, footwear, medical prostheses, and more, all can be customizable to the individual. Our smart homes enable us to tailor the environment around us by simply speaking. Even our food is becoming personalized, using Sifter's shop-by-diet guidance in the store, or the many restaurants that will customize a meal based on food allergies, dietary requirements, lifestyle preferences, and more.

And there are implications for the retail industry.

Certainly there are implications around marketing. In the Age of I, consumers increasingly expect promotions and savings to be relevant; they don't expect or value the mass promotions of the past century. Each individual customer and household is effectively saying, "help me save money on the products I want to buy, when I want to purchase them." They aren't saying they want them when the brand may want to promote them. It is this trend that makes what Kevin Stafford, VP of Marketing at KVAT Food City, and other retailers like Weis Markets, Coborns, and others, are doing to provide personalized savings to each shopper so powerful.

And it is this same philosophy that retailers can leverage in using tools like Replenium's subscription capability, to help shoppers to save money on the products they want while capturing more consistent share-of-wallet for the retailer.

But the implications for retail go beyond personalizing promotions and marketing content. Customer recognition is one of the most powerful tools there is, universally valued, and something retailers—especially larger retailers—simply don't do.

I wrote about M&M Meat Shops, the Canadian retailer, earlier in the book. Mac and the team at M&M strongly believed that their franchisees had a massive opportunity to grow loyalty through customer recognition. A year or so after they launched their program, the head office coordinated a program whereby each store owner delivered a bouquet of roses to their top ten customers on Valentine's Day. The initiative had the entire country of Canada talking, creating a buzz around M&M Meat Shops. That's the power of customer recognition.

While AI and machine learning capabilities can efficiently scale personalization, there remains the power of human interaction. I believe retailers that use technology and innovative capabilities to combine the two will be the winners in the time ahead.

Retail in the Age of I captures my retail philosophy: Retail is truly about serving the individual customer and using technology to deliver at scale the kind of personalized service that my grandfather provided decades ago.

Time for a Change

Insights to true customer value and behavior coming out of our loyalty program nearly thirty years ago only reinforced the belief that the customer truly was the most important part of the business. It is bewildering to me the number of retailers who continue to believe their revenue comes from the box of cereal on the shelf and not the customer opening their wallet (or waving their palm over the payment terminal!) to purchase it.

Supermarket operators, in particular, should take note.

According to the Department of Agriculture, supermarkets' share of American's total food spending was around 37 percent in 1997. By 2022 it was down to around 25 percent, with super centers and warehouse clubs gaining a good part of that, while restaurants, including fast food, grew their share of household food spending to about 37.5 percent.

Stolen Lunch

Share of total food spending

- Grocery stores
- Full-service restaurants
- Limited-service restaurants
- Retail stores and vending
- Warehouse clubs and supercenters
- Mail order and home delivery

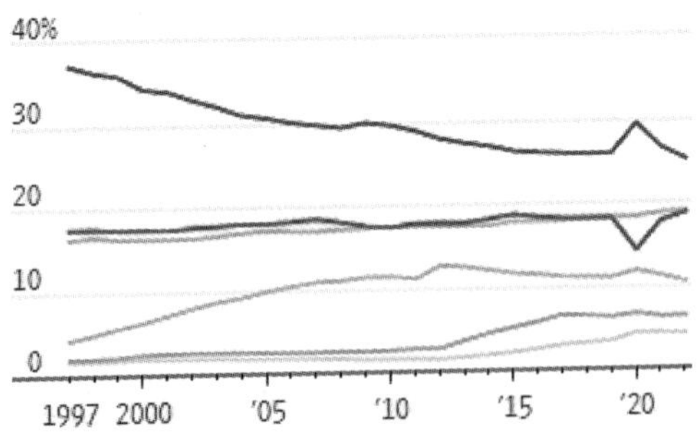

Source: U.S. Department of Agriculture

"Supermarkets' share is most endangered in the so-called center aisle, where many competitors sell the same brands of toilet paper, cleaning materials and cereals, often at lower prices. They simply have more advantaged business models on those types of repeat purchases. Costco, for example, has a highly efficient operating model and leans on membership fees and sales volume rather than retail margins. Mass merchants such as Walmart and Target sell a mix of products—such as apparel and home products—with higher margins to compensate for low profits on consumables. Discount grocers, meanwhile, have smaller formats and a narrower selection of products, reducing overhead. And then there is the internet: It is difficult to beat the convenience of having repeat purchases delivered directly to consumers' doorsteps. Amazon has made these purchases stickier by offering discounts for households that opt to 'subscribe and save.'"*

Despite this, traditional supermarket retailers continue to rely upon the same business model that they have for decades, a heavy reliance on CPG brand manufacturer trade funds. "So-called trade funding can account for more than a half of supermarkets' operating profits, according to Matthew Hamory, partner at consulting firm AlixPartners. Relying on that source of profit could start to work against them. Manufacturers will naturally want to allocate more of those dollars to retailers growing the most sales volume, which recently have been big-box mass merchants like Walmart and warehouse clubs like Costco, according to Scott Mushkin, chief executive of industry research firm at R5 Capital."†

As we venture forth into the exponential world and all its disruption, supermarket operators—indeed, all retailers—have an opportunity to remake themselves, and to reclaim their customer heritage.

Awe-Inspiring Shopping Experiences

I started out this book talking about the power of inspired vision. Technology today enables the largest, and the smallest, retailers to create and realize their unique vision of the future of retail . . . but only if retailers are willing to change.

---

* Jinjoo Lee, "Supermarkets Are Losing This Food Fight," *Wall Street Journal*, July 21, 2023, https://www.wsj.com/articles/supermarkets-competition-costco-walmart-aldi-4f3c0d0c.

† Ibid.

We took SmartShop, the personalized marketing capability we had built, live in 2005. I will never forget an older customer stopping me that first day. She had come in the store, picked up a dozen or two products in her cart, and then saw the SmartShop kiosk. Using her finger scan (using Pay By Touch's biometric tech talked about earlier), she printed out her list of personalized offers . . . and was awestruck to find several products she had already put in her cart on her printout giving her additional, personalized savings. She told me it was amazing, like someone had read her mind, not only finding some of the products she had already picked up on her list but receiving additional savings.

The awe that shopper experienced spread fast across other shoppers. In the following months, we saw a growing number of shoppers entering the store, getting a cart or basket, and then going to the kiosk to see their personalized offers. We saw shoppers dragging their relatives, friends, and neighbors to the kiosks, saying, "You've got to see this."

That shopper's experience was the embodiment of my vision to bring personalized savings to retail. The journey began with inspiration, the inspiration becoming a vision, and then committed action to surmount the obstacles to bring the vision to life.

Becoming bionic gives traditional retailers an opportunity to regain the customer focus the massive industry was built on. Leveraging AI and machine learning applications to provide personalized savings to each shopper. Giving the shopper control over their data. Using AR to make the shopping experience easier, more efficient, and more exciting. Bionic powered awe.

Imagine a shopper approaching the cheese counter, the shopper's name—let's call him Mark— appearing in the cheesemonger's AR smart glasses, triggered by the retailer's app on Mark's smartphone where he has agreed to share data with the retailer. In addition to Mark's name, his past cheese purchases, along with wine preferences, called up instantly and displayed, enabling the cheesemonger to guide Mark to new cheeses that he might enjoy. Experiences like this—highly curated, relevant, and personalized interactions—building relationships and loyalty. And this capability is just around the corner.

Zach Bello relates his experience with Spotify to highlight the use of technology in transforming markets and in building those long-time relationships. As Zach puts it, "Before Spotify, you'd walk into a music store to browse thousands of tapes, CDs in static categorization. Today, in seconds, you can browse essentially the worlds entire catalog of music and discover music in categories you didn't even know existed. Plus, from a loyalty perspective, I'd be hard pressed to ever leave Spotify because of how well they know my tastes in music." Today, Spotify has over 500 million active users and over 100 million songs.

Regardless of what type of retail you're in, think about how to use technology to create that kind of user experience, an experience that has your customers saying, "I'd be hard pressed to ever leave [your business] because of how well they know me." These types of technologies can be leveraged in a way that strengthens your business and brand but does not deplete or overrule the human connections and appreciation happening in your locations.

Your Mission, Should You Decide to Accept It

In the TV series *The Six Million Dollar Man*, Steve Austin becomes a secret agent, using his newfound superhuman abilities to carry out his mission and win the day. Likewise, the bionic retailer has a noble mission: To make retail better, stronger, and faster, while returning its focus to the individual customer.

As the world of tech-fueled innovation moves faster and faster, retail leaders must be thinking further ahead than just becoming an omnichannel retailer. The future is about more than simply plugging in new technological capabilities. It is about fusing together physical automation and sophisticated artificial intelligence integrated through a digital nervous system into our human organizations to create a new retail species: Bionic Retail.™

And as we go forward, retailers have an incredible opportunity to become stewards to the future in partnership with their customers and their associates, and in doing so, not only build shopper loyalty, but lifelong relationships.

But only if retailers understand that they are now operating in a world unlike any before, a world past the inflection point on technology's exponential growth curve.

My intention in writing this book is to help retailers understand they must shift focus, look past the new shiny innovations coming into the market almost daily, and understand that the organizational structures, decision making processes, and traditional retail operating and business models, are all increasingly misaligned with the future as we accelerate up and out the curve.

Most of all, I want to share the blueprint that has come out of my journey, my experiences, and my work with retailers and technology providers around the world. Added to that blueprint are countless discussions with my son Sterling and daughter Haviland, bringing together new, generationally different, viewpoints, knowledge, and experience.

The power of an inspired vision, fueled by committed action, and guided by a doctrine that embeds needed values and beliefs in a company's organization and culture, is incredibly powerful. And it is the path for readers to create the future of retail.

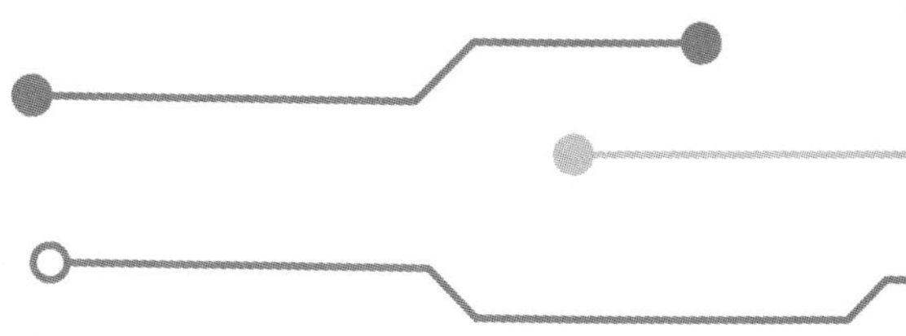

ABOUT THE AUTHOR

Gary Hawkins has been described as a visionary and retail industry treasure, whose deep knowledge of the industry combines with operational experience and technology savvy to provide a perspective that is completely unique.

His Odyssean journey across retail began three decades ago, driven by a vision to use data and technology to transform retail, returning the massive industry to its customer birthright. That inspiration in turn gave rise to successive waves of innovation that continue today. Along the way he's decimated countless industry practices on a relentless march into the future.

That work has taken him to five different continents, helping premier retailers and CPG brand manufacturers discover and understand new technologies while helping the creators of new capabilities understand the massive and dynamic consumer goods retail industry.

Today, drawing from his experiences and learning, Hawkins and his team have created a powerful blueprint to help the retail industry thrive in an exponential world. Retailers have a unique opportunity to set aside outdated beliefs and use accelerating innovation and never-before-possible capabilities to create a powerful vision for retail's future.

Hawkins leverages his expansive industry view, and early insight into disruptive technology, as a sought-after keynote speaker at conferences in the U.S. and around the world. Hawkins is the author of *Building the Customer Specific Retail Enterprise*; *Customer Intelligence*; *Retail in the Age of I,* and *Bionic Retail,* along with myriad articles and white papers. Hawkins lives in Colorado with his wife Heather, and Remington, their Bernese Mountain dog.

GARYHAWKINS.INFO